The Politics of Rural Reform in China

Based on a treasure trove of information collected through fieldwork interviews and painstaking documentary research through the Chinese and Western language presses, this book analyzes one of the most important reforms implemented in China over the past decade – the Rural Tax and Fee Reform, also known as the "Third Revolution in the Countryside". The aim of the reform to improve social stability in rural China, which has become increasingly shaken by peasant protests, many of them large-scale and violent. By examining the gap between the intentions of the reform and the eventual outcomes, Göbel provides new insights into the nature of intergovernmental relations in China and highlights the ways in which the relationship between the state and the rural populace has fundamentally changed forever.

The Politics of Rural Reform in China will appeal to students and scholars of Chinese politics, governance and development studies.

Christian Göbel is a researcher of Political Science at the Centre for East and South-East Asian Studies, Lund University, where he also teaches Comparative Politics.

Chinese Worlds

Chinese Worlds publishes high-quality scholarship, research monographs and source collections on Chinese history and society. 'Worlds' signals the diversity of China, the cycles of unity and division through which China's modern history has passed, and recent research trends toward regional studies and local issues. It also signals that Chineseness is not contained within borders: ethnic migrant communities overseas are also 'Chinese worlds'.

The Literary Fields of Twentieth-Century China
Edited by Michel Hockx

Chinese Business in Malaysia
Accumulation, ascendance, accommodation
Edmund Terence Gomez

Internal and International Migration
Chinese perspectives
Edited by Frank N. Pieke and Hein Mallee

Village Inc.
Chinese rural society in the 1990s
Edited by Flemming Christiansen and Zhang Junzuo

Chen Duxiu's Last Articles and Letters, 1937–1942
Edited and translated by Gregor Benton

Encyclopedia of the Chinese Overseas
Edited by Lynn Pan

New Fourth Army
Communist resistance along the Yangtze and the Huai, 1938–1941
Gregor Benton

A Road is Made
Communism in Shanghai, 1920–1927
Steve Smith

The Bolsheviks and the Chinese Revolution, 1919–1927
Alexander Pantsov

Chinas Unlimited
Gregory Lee

Friend of China
The Myth of Rewi Alley
Anne-Marie Brady

Birth Control in China, 1949–2000
Population policy and demographic development
Thomas Scharping

Chinatown, Europe
An exploration of overseas Chinese identity in the 1990s
Flemming Christiansen

Financing China's Rural Enterprises
Jun Li

Confucian Capitalism
Souchou Yao

Chinese Business in the Making of a Malay State, 1882–1941
Kedah and Penang
Wu Xiao An

Chinese Enterprise, Transnationalism and Identity
Edited by Edmund Terence Gomez and Hsin-Huang Michael Hsiao

Diasporic Chinese Ventures
The life and work of Wang Gungwu
Gregor Benton and Hong Liu

Intellectuals in Revolutionary China, 1921–1949
Leaders, heroes and sophisticates
Hung-yok Ip

Migration, Ethnic Relations and Chinese Business
Kwok Bun Chan

Chinese Identities, Ethnicity and Cosmopolitanism
Kwok Bun Chan

Chinese Ethnic Business
Global and local perspectives
Edited by Eric Fong and Chiu Luk

Chinese Transnational Networks
Edited by Tan Chee-Beng

Chinese Migrants and Internationalism
Forgotten histories, 1917–1945
Gregor Benton

Chinese in Eastern Europe and Russia
A middleman minority in a transnational era
Nyíri Pál

Chinese Entrepreneurship in a Global Era
Raymond Sin-wok Wong

The Politics of Rural Reform in China
State policy and village predicament in the early 2000s
Christian Göbel

The Politics of Rural Reform in China

State policy and village predicament in the early 2000s

Christian Göbel

Routledge
Taylor & Francis Group
LONDON AND NEW YORK

First published 2010
by Routledge
2 Park Square, Milton Park, Abingdon, Oxon OX14 4RN

Simultaneously published in the USA and Canada
by Routledge
270 Madison Ave, New York, NY 10016

Routledge is an imprint of the Taylor & Francis Group, an informa business

Typeset in 10/12 pt Times NR MT by
Graphicraft Limited, Hong Kong
Printed and bound in Great Britain by
CPI Antony Rowe, Chippenham, Wiltshire

British Library Cataloguing in Publication Data
A catalogue record for this book is available from the British Library

Library of Congress Cataloging in Publication Data
Göbel, Christian, 1973 June 25-
The politics of rural reform in China : state policy and village predicament in the early 2000s / Christian Göbel.
p. cm. – (Chinese worlds)
Includes bibliographical references and index.
1. Rural development–China. 2. Rural development–Government policy–China. 3. China–Economic policy. I. Title.
HN740.Z9C637612 2010
338.951009173′4–dc22

2009050245

ISBN: 978-0-415-55935-5 (hbk)
ISBN: 978-0-203-84976-7 (ebk)

To Heike-Alice
for bringing joy, wisdom and inspiration to my life

Contents

Illustrations

Tables

Acknowledgements

Many people have contributed material and ideas to this publication, made the project possible, my fieldwork in China memorable, or supported me at home. I would first of all like to thank the German Academic Exchange Service (DAAD), whose generous financial support made eighteen weeks of fieldwork in China possible. I am further indebted to Chang Xinxin for opening doors that would otherwise have stayed closed; to Jean Hung for academic 'match-making'; to ZJP, ZJJ and ZNN for political 'match-making'; to Zhu Shiqun and Li Yuanxing for supporting my research in Anhui; to Graeme Smith for a lush mountain retreat in 'Benghai County'; to Flemming Christiansen for helping me find the right words; to Zhu Qinfang and Franz Li for being the perfect hosts; to Hans-Joachim Wenzel for hundreds of newspaper clippings; to David Kühn, Gunter Schubert and the anonymous referee engaged by Routledge for extensive comments on the draft of this study; to Andreas Fulda for housing me in Beijing and for his steady flow of comments and suggestions; and of course to all my interviewees for their trust and the information without which the book could not have been written. They are listed in the appendix, but must remain anonymous. In particular, I am indebted to W.K.N., L.Q. and Z.X. for their friendship and for providing me with unforgettable memories in rural China. I owe a special debt of gratitude to Thomas Heberer for being a patient listener, an avid teacher and a constant source of learning, encouragement and personal support. Finally, I wish to thank Sonja van Leeuven and Stephanie Rogers at Routledge for their guidance in publishing the manuscript, and Leanne Hinves and Kathy Auger for their assistance in the copy-editing process.

Abbreviations

ABC	Agricultural Bank of China (Zhongguo nongye yinhang)
ATNJ	*Anhui tongji nianjian* (Statistical Yearbook of Anhui Province)
CASS	Chinese Academy of Social Sciences (Zhongguo shehui kexueyuan)
CC	Central Committee
CPC	Communist Party of China
CZNJ	*Zhongguo caizheng nianjian* (Fiscal Yearbook of China)
DRF	Development Research Foundation (Guowuyuan fazhan yanjiu jijinhui)
DRI	Development Research Institute of the State Council (Guowuyuan fazhan yanjiu zhongxin)
EBR	Extra-budgetary revenue
ESR	Extra-systemic revenue
FBIS	Foreign Broadcasting and Information Service
GDP	Gross domestic product
HRS	Household Responsibility System
IOU	'I owe you' (written acknowledgement of a liability)
MoA	Ministry of Agriculture
MoF	Ministry of Finance
n.a.	not available
NI	New Institutionalism
NPC	National People's Congress
PAF	Principal–agent framework
PCI	Per capita income
PD	People's Daily, the *Renmin ribao*
PPCI	Peasant per capita income
PRC	People's Republic of China
RMB	Renminbi (Chinese currency)
RTFR	Rural Tax and Fee Reform
SOE	State-owned enterprises
TVE	Township and village enterprises
WTO	World Trade Organization
ZTNJ	*Zhongguo tongji nianjian* (Statistical Yearbook of China)

1 The peasant burden and the dynamics of rural reform and development

Unjust exactions from the Chinese peasantry have been a driving force of China's history. Most dynasties were founded on the pledge to improve the lot of China's farmers, and began with good relations between those governing and those governed. Gradually, however, the administrative apparatus grew, and corruption set in. In order to satisfy their ever-increasing needs, the bureaucrats extracted more and more resources from the Chinese peasantry. Eventually, a tipping point was reached. Infuriated peasants led by disgruntled bureaucrats either brought down the regime themselves, or destabilized it just enough for contenders to the throne to take over. The new emperor drastically simplified the fiscal system, and the process started all over again.

This 'dynastic cycle' is a very strong image in Chinese political thought.[1] Given that the People's Republic of China was also founded on the pledge to liberate China's farmers from feudalism and oppression, the current leadership has every reason to be concerned about China's stability. For a country that claims to be socialist, income inequality is alarmingly high. China's Gini index stood at 46.9 in 2007, on a par with countries such as Rwanda and Mexico.[2] While the richest 10 per cent earned more than one-third of all income, the poorest 10 per cent have to get by with just over 1 per cent. As a consequence, social tensions have increased. According to official figures, the number of 'mass incidents' rose 28 per cent between 2003 and 2004. Besides rural protests, another worrisome phenomenon was observed in the beginning of the twenty-first century. As a result of falling market prices due to large-scale grain speculation,[3] grain output dropped from 508.4 million tons in 1999 to 430.7 million tons in 2003.[4] This alerted the central government that the country's policy of self-sufficient food production could be undermined in the long run.[5]

It is no coincidence that the beginning at the twenty-first century saw a shift in development priorities in the central government. The exclusive focus on economic growth came to be replaced by a pledge to sustainable development, which also entailed devising solutions for China's manifold rural problems. In 2006, the central government started to devote large subsidies to an ambitious and costly rural development programme termed the 'New

Socialist Countryside' (*shehui zhuyi xin nongcun*). Part of the programme is the promise to connect all villages to the outside by tar roads, to provide medical insurance for all farmers by 2020, to extent the minimum living allowance (*dibao*) policy from the cities to the rural areas, and to improve agricultural mechanization.[6] One of the key issues at the subsequent Seventeenth National Congress of the Communist Party of China in October 2007 was including in the Party Constitution the commitment to 'balance' urban and rural development.[7] Urban–rural integration was also at the heart of the Third Plenary Session of the Seventeenth NPC in October 2008, where the Party's pledge to 'rural reform and development' was renewed by promising to double peasant income by 2020 and to further improve the rural infrastructure.[8] Statements made by high-level leaders indicate that more attempts to remove the urban–rural bias will follow. For example, the Chinese President, Hu Jintao, indicated in early October 2008 that a land transaction market would be created, allowing peasants to freely trade, rent and even mortgage their land-use rights.[9] Reforms to China's residence registration (*hukou*) system that would make it easier for peasants to take up residence in the cities are also envisioned.

Precluding – and to some extent shaping – this renewed focus on urban–rural integration was the Rural Tax and Fee Reform (RTFR), a set of reform policies that had a major impact on rural development in China. Given its importance, it is surprising that it has so far received little scholarly attention outside China. The RTFR was designed to simultaneously reduce the fiscal burden on the peasants and increase their income, improve local governance, and develop the countryside. It was formally implemented in 2000, and the name belies the scope of institutional change it entailed. Perhaps in order to make up for the rather bland name, officials called the reform euphemistically the 'Third Revolution in the Countryside' (*di san ci nongcun geming*).[10]

The Rural Tax and Fee Reform

Unlike earlier approaches to easing the fiscal burden of the peasants, which had all failed, the RTFR combined the simplification and formalization of local revenue collection and redistribution with far-reaching institutional reforms at the local level.[11] As one document put it: 'The Rural Tax and Fee Reform not only implies the regulation of the distributive relationship between national and peasant income, it is a large transformation of the village superstructure.'[12] The reform aimed at not only restructuring, but building anew local-level government structures with the ultimate aim to improve state distributive performance and thus to defuse the critical situation brought about by what is commonly known as the 'three rural problems' (*san nong wenti*).[13] In addition, the formative document of the reform makes clear that the ultimate aim of the RTFR was not only the improvement of rural stability, but also the strengthening of the national economy: by transferring a larger part of the GDP to the peasants, the purchasing power of the majority of China's population would be increased and agricultural productivity enhanced. Thus, from the beginning, the 'main reform' (*zhuti gaige*) of abolishing all rural taxes, fees and levies except for the agricultural tax was accompanied by an imperative to implement 'complementary reforms' (*peitao gaige*)[14] in order to improve local governance.[15] First of all, local governments were required to streamline, shed superfluous employees, privatize services, merge village schools, and reduce the number of teachers. Second, the 'fiscal administration' at the county and township levels was to be improved. Third, a 'burden supervision system' was to be established.[16] The aim of the latter measures was to make the reform sustainable by decentralizing local prerogatives, but also to save administrative costs and thereby compensate for local-level revenue losses resulting from the tax and fee reduction.

The RTFR developed in several stages. Between 1993 and 2000, more than fifty counties in seven provinces experimented with alternative tax and fee models. In mid-2000, the central government presented a compulsory reform model, which at first was only implemented in Anhui Province. This policy mandated the abolition of most regular and irregular fees at the price of a tripled agricultural tax. At first, the RTFR was celebrated as a great success. When it was tested in Anhui, the peasant burden was reported to have decreased almost 38 per cent.[17] At the national level, average burden reduction even reached 45.8 per cent after the reform's extension in 2003.[18] However, a large variation underlies these figures. For example, average per capita peasant burden in Anhui's Dangshan County was reduced 75.6 per cent between 1999 and 2003, but the corresponding figure for Dangtu County was only 3.9 per cent.[19] With the abolition of the high tax on agricultural cash crops (*nongye techan shui*), per capita burden was reduced more than 60 per cent during that time in the tea-growing counties of southern and western Anhui. In many central Anhui counties, however, the burden reduction rate lay significantly below the national average.

Such a high variation would not have been problematic if the poorer peasants had benefited from the larger reductions. As several studies show, however, this was not the case. The poorer a locality was, the more the peasants actually suffered from the impacts of the reform.[20] That reform implementation was not unproblematic is also revealed by the increase in complaints that were lodged against the government of Anhui Province (see Chapter 6).

As John James Kennedy has shown, the peasants in poorer regions were not only affected by continued fee-taking, which was the main source of these complaints, but also by the negative impact the reform had on the provision of medical services and basic education.[21] In fact, the beneficial effects of burden reduction were offset by the higher expenses these peasants now had to shoulder in order to be able to afford medical treatment and sending their children to school.

It is safe to say that the RTFR did not bring about the desired results,[22] and that it was the failure of the initial policy design which prompted policy-makers to announce in early 2004, only one year after the policy had been implemented nationwide, that the agricultural tax would also be abolished. In fact, the central-level document that heralded this step conceded failure by stating that 'not all problems in the RTFR pilot regions have been solved. The peasant burden is still rather heavy, and complementary reform implementation lags behind. The fundament of the peasant burden reduction is not yet consolidated.'[23]

What followed the abolition of the agricultural tax was in many ways the opposite of what had been intended in the initial reform design. Whereas the aim had been to merely lighten the burden of the peasants, two thirds of China's population were almost totally freed of having to contribute to financing the state. Furthermore, the need to keep village- and township-level administrations operating on a regular basis had been recognized as one of the 'basic principles' of the reform, but as a result of heavy revenue losses, this principle could not be upheld in many places.

In a related manner, the centre planned not to get involved in financing this reform, but ended up footing the bill: In Anhui Province, for example, the ratio by which expenditures exceded revenues climbed from 40 per cent in the pre-reform year to almost 288 per cent in 2006, the overwhelming part of the deficit being financed by transfer payments form the central government.[24] Another ironic reform outcome is that the number of administrative cadres increased between 2002 and 2005 after having stagnated for ten years despite the fact that a chief aim of the reform was to downsize the cadre workforce.[25] A possible reason for this is that county governments added personnel in order to facilitate the implementation of the RTFR.

It is questionable if the adjustments in 2004 were indeed, as the central government portrays is, the 'second step' of the RTFR.[26] It seems more likely that, as Linda Chelan Li has put it, a series of unintended consequences resulting from strategic interaction in a multi-level system was depicted in retrospect as 'a succession of co-ordinated central measures with coherent objectives.'[27] In any case, except for the fact that the peasant burden was indeed reduced significantly after the abolition of the agricultural tax (albeit at the price of reduced education and health care services), the reform did not develop as planned.

Explaining the policy outcomes

The RTFR has brought significant changes to the Chinese polity that are worthy of being researched in their own respect, but also in their function as indicators for the changing nature of intergovernmental relationships and in their potential to affect China's stability and long-term development. While Chinese scholars have recognized very early the significance of the reform, its theoretical and practical implications remain under-researched.

This book provides explanations for the unintended outcomes of the RTFR by tracing its development from the very beginning in the late 1980s up to its conclusion in 2006. It does so by shedding light on the intricate relationship between political structures and agency in the black box of the policy process. In other words, it tackles the question if the implementation gap was mainly a result of path dependence or independent agency.

As just shown, the purpose of the RTFR was not only to reduce the peasant burden, but also to change the local-level institutional and organizational framework within which policies are implemented. Given that the RTFR itself is such a policy, this is an intriguing situation for the practitioner and the scholar alike. Institutional reforms reshape structures of authority and power. If actors are at once involved in the reform process and affected by its outcome, their stakes in shaping this process are high. Accordingly, they will strive not only to influence the implementation of the policy. Much earlier, they will seek to ensure that its design converges with their aims. The more they are able to influence the interpretation of the problem, the formulation of the policy, and its implementation, the more they can expect that the outcomes of the reform meet their expectations. For this reason, institutional reforms always walk the tightrope between solving an 'objective' problem and balancing the legitimate interests of the parties which are at once involved and affected.[28] As we know from the literature on democratic transitions,[29] the outcome of such self-referential institution-building is highly uncertain, especially if no rules or organizations are created to bridge the transition from the old to the new system. Such situations are highly contingent on the strength of the actors that seek to maintain or change these rules, but also on the resilience of the existing rules of the political game. Both ideas, the role of actors and the role of institutions in shaping reform outcomes, require some elaboration.

Principals and agents

The relationship between the policy designers in the centre and policy implementers at the local level is an asymmetrical one. It is best understood as one of a 'principal' and his 'agents'. Put very simply and applied to the research question of how central–local relations affected the outcome of the RTFR, the central government (principal) wants the local governments (agents) to provide public services and put into practice specific policies. In exchange, the agent receives a gratification from the principal, be it in the form of a salary, a reward, or sometimes even only praise. However, although principal and agent depend upon each other, they might have different agendas, and the agent might be quite unwilling to do the principal's bid. As the following sections will show, there is a battery of instruments the principal can use to have his goals achieved. The still most common instrument (not only in China) is the use of negative sanctions. The agent is punished if he deviates from the principal's policy. But how will the latter know? The ancient Chinese witticism that 'heaven is high and the emperor far away' (*tian gao, shangdi yuan*) shows that local politicians have grasped a central feature of the principal–agent relationship: that information asymmetry due to distances in time and space, or even information overload, can be quite beneficial to the agent. He might promise something, but deliver something of inferior value (shirking) or even thwart the principal's policy (sabotage)[30] and pretend that he has done a beautiful job. When the principal finally finds out (if he ever does), it might be too late to reverse the damage done. Most scholars would agree that this is exactly what characterized Chinese politics throughout history: Chinese and foreign scholars alike all agree that the negative outcomes discussed above can be traced back to corrupt and selfish cadres over whom the centre has lost control.[31] In the case of the RTFR, for example, illegal fee-taking continued unabated in several regions,[32] local governments discovered legal loopholes through which to further extract moneys and goods from the peasants,[33] and the design of the new agricultural tax proved highly unfair and subject to easy manipulation.[34]

However, principal–agent relationships do not need to be antagonistic: Linda Chelan Li has provided arguments against such a simplistic view,[35] and has convincingly made the case for replacing the traditional 'zero-sum framework of spatial politics'[36] with a 'non-dualistic' approach emphasizing 'the coexistence of central and local power in a diffuse, complex decision-making process'. The central and local actors are conceptualized as 'co-participants' in the reform process who 'interact and support one another', yet are 'analytically distinct with obvious boundaries'. More concretely, the centre's role is to provide a guiding framework for both the parameters of the desired policy and local-level behaviour, while the sub-national actors' task lies in implementing and improving national policy.[37] Conflicts are not antagonistic, since localities contravene the initial policy in order to correct it. By trying out innovative solutions to existing problems, they can even act as reformers.[38] Using the reform of the tax on agricultural special products as an example, she illustrates that workable policies emerging from below are likely to be adapted even if they contradict existing guidelines.[39]

The two positions just presented are not mutually exclusive. Obviously, principal and agents cooperate in some policies, but not in others. What are the determinants of cooperation? According to Sebastian Heilmann, 'experimental programmes that were seen as conducive to economic growth and generating income within a particular jurisdiction were vigorously pursued by local officials'.[40] On the other hand, '[l]imitations of experimentation are obvious in improving the provision of social and public goods, such as access to basic health care and credible protection of land tenure rights in rural areas or environmental protection'.[41]

Local officials, this view holds, are guided by short-term interests and only innovate and cooperate if they can reap tangible financial benefits, a point also made by Jean Oi.[42] The RTFR clearly did not offer any opportunities to collect dividends, so its failure seems to have been predestined. Although highly plausible, this theory cannot explain the bottom-up character of the reform, which had been preceded by local experiments for almost a decade. Throughout the 1990s, the centre gave only lukewarm support to the policy, and when it finally did become a national policy in the late 1990s, it was marred by 'elite competition and bureaucratic boundaries' at the central level. According to Linda Chelan Li, this resulted in insufficient reform preparation and sloppy implementation,[43] and the instructions the localities received were often not very clear and sometimes even contradictory. In essence, responsible for the implementation gaps in the RTFR was not an agency problem, but rather a 'principality problem'! While the outcome is similar, the difference in disincentives is very subtle: in the case of the agency problem, the agent *does not want* to implement the principal's policies, but in the case of the 'principality problem', it receives unclear orders. In fact, it might not be clear even to central-level decision-makers what these policies are or should be. Hence, the reform's failure cannot be blamed on central or local actors alone, which leads Linda Chelan Li to argue that the reform outcome is the result of 'contingencies and unintended consequences' that eschew simple theoretization.

While this would mean that from studying the RTFR we can learn only about the RTFR, this book argues that a careful analysis of the reform will provide valuable insights that transcend this specific policy. In order to achieve this, however, it is necessary to look not only at actors, but also at the political rules that structure their interaction. For example, a volume composed by an editorial body of the State Development Planning Commission (*guojia fazhan jihua weiyuanhui*) argues that the root of the peasant burden lies in the 'serious lag' of administrative reform, the incompleteness of the 1994 tax reform, the ineffectiveness and inefficiency of agricultural taxation, the fact that it is mainly the wealthier grassroots governments who profit from off-budget levying,[44] and the lack of organizations promoting peasants' interests.[45] Although the RTFR partly addressed these issues, the authors of the central government-sponsored study argue that the local-level economic and administrative structures, local government indebtedness, unbalanced development and the lack of transparency constrained the reform.[46]

These observations fall into a more institutionalist line of scholarship which has analysed the impact of decentralization or (re)centralization measures on the outcomes of certain reform policies. Decentralization has long been understood as a powerful tool to enhance local government efficiency, accountability, responsiveness and transparency.[47] It is easy to see that, theoretically, moving the centre of service provision closer to the affected constituency will both enable and pressure governments to provide better services and reduce the transaction costs of public service provision.[48] In a related manner, it has been argued that the decentralization of economic decision-making is a recipe for economic growth, a claim that seems to be readily validated by the Chinese and Indian experiences.[49] Finally, political science has tended to assume that '[decentralization] can promote a sense of autonomy in citizens, enhance social order by promoting the legitimacy of the state, and limit pressures for separatism . . .'. As Paul Hutchcroft points out, however, 'many of these initiatives often seem to rest more on faith than on strong conceptual foundations or careful analysis of historical and comparative experiences.'[50] Newer studies support such a cautious view and stress that the transfer of rights and responsibilities from the central to the local level can lead to sub-optimal outcomes especially in developing countries with weak political centres.[51]

Competition under hierarchy in China

This seems to be true for not only for the RTFR, but also for the burden problem that made this reform necessary. The following sections therefore raise some critical issues regarding the impact of decentralization on economic development, the peasant burden and rural reform politics. As regards the impact of decentralization on economic development, China is often regarded as a showcase of success. However, this view exaggerates the degree of administrative autonomy local governments enjoyed before the RTFR was implemented, and it overemphasizes the virtues, and neglects the negative consequences, of decentralization in China. As for the depth of decentralization in China, Chapter 3 will show that administrative decentralization after 1978 did not take the most extensive form of 'devolution' (complete autonomy), but rather of 'delegation' (autonomy in planning and management, but accountability for the outcomes). More specifically, the central government set mandatory targets and merely let the localities work out the details. For example, economic growth was (and still is) the single most important indicator for evaluating cadre performance, so local cadres were not free to neglect economic growth. The same was true for various other administrative fields as diverse as 'law enforcement and security, purchase of insurance policies, sports competitions, sanitation, primary health care services, and the building of schools, agriculture and healthcare centres, old-age homes, or police stations.'[52] Local governments were given limited decision-making powers within a corridor constrained by the centre's achievement standards. This led to the negative consequences just mentioned: limited decentralization spurred development where certain infrastructural conditions were met, but proved detrimental for more backward regions. Frequently, poor local governments were required to spend more than they could afford to fulfil these targets. Since they were unable to tap communal enterprises for taxes and off-budget revenue, they were eventually forced to exploit the peasants.

However, administrative delegation alone does not explain the existence of a heavy peasant burden. As Andrew MacIntyre has shown in a different context, altering the division of rights and responsibilities is a necessary, but not sufficient precondition for changed political or economic outcomes.[53]

Centralization and decentralization merely delineate the division of rule-making capacities or, put differently, who has the authority to do what. As explained above, decentralization alone does not explain reform failure. What has also to be considered is the equally important question of how this authority is put into practice.[54] Although dozens of instruments to steer or influence the making and implementation of policies have been identified,[55] they can be neatly grouped into three ideal types of exerting influence in intra-governmental relationships. As Kenis and Schneider point out, '[m]arket, hierarchy and network are the most important forms of reference'.[56] Indeed, moving from one to another of these three ideal types forms the base of most administrative reform proposals in the contemporary world.[57]

As is the case with centralization and decentralization, Christopher Hood has shown that all three of these types are associated with particular strengths and weaknesses. Hierarchic structures provide control, expertise and predictability, and careful planning can help prevent the occurrence of 'public management disasters'.[58] On the other hand, however, this feature makes them inflexible and therefore particularly vulnerable to unanticipated events. In addition, functioning hierarchies need intact chains of command, i.e. agents who do not abuse their position to realize particularistic goals that are incompatible with those of their organization.[59] In contrast, competitive structures are more flexible and need less supervision, because actors are evaluated not primarily according to how well they stick to existing rules of procedure, but according to the outcomes they deliver. If agents are made to compete, it is not the threat of coercive force which structures their incentives, but the expectation of rewards if they perform well. Markets, however, stress individualism and neglect the collective benefit. Efficiency is deemed more important than equality, which hinders co-operation and can further corruption.[60] Egalitarian networks, finally, are neither prone to agency problems nor to inequality, because they rely on democracy, empowerment and group decisions. Such decisions, however, tend to come about slowly, and the egalitarians' unwillingness to accept a higher authority to break deadlocks can lead to unresolved feuds which can break the community apart.[61]

While the structural dimension of intergovernmental relations has received much scholarly attention, this process dimension of policy-making and implementation is still remarkably under-researched.[62] This study shows that a particular type of a spatial division of labour, i.e. delegation, facilitates a particular type of policy implementation steering, i.e. competition under hierarchy, which bears large responsibility for the problems just discussed. A mixture of two of the above ideal types characterizes this style, namely the hierarchical prescription of more or less specific outcomes and the stimulation of competition to achieve or even exceed these targets. More specifically, only abstract achievement standards are formulated (called 'unified national policy'), but no guidelines on how to achieve them (called 'dispersed policy-making'). This method is the result of a paradox that has characterized central–local relations since Reform and Opening and even before. To prevent the localities from distorting central policies or not implementing them at all, the application of 'hierarchical' instruments of regulation, supervision and punishment is deemed necessary. On the other hand, however, policies cannot be implemented uniformly in China's nearly 50,000, vastly different township-level entities if gross unfairness is to be avoided. Hence, the localities must be counted upon to work out the implementation details. At first, it might seem astonishing that competition is a major feature in Chinese politics, but it is not so surprising if one considers the role of models to be emulated in Chinese administrative history. After all, the excessive competition stimulated by setting grain quota standards to be achieved or even exceeded is one of the chief reasons why China's Great Leap Forward (1958–61) turned out so disastrous.[63] In fact, the setting and even increasing of unrealistic quotas today does not differ much from policy implementation during the planned economy era.

As the Chinese case shows, competition under hierarchy can spur economic growth under favourable conditions, but can also prove counterproductive in a system where the affected agents have no exit option. In a free market, a supplier can always choose not to produce a good if the customer dictates a ridiculous price. Within an administration, however, the agent will lose his job if he does not at least pretend to comply. Cooperative elements are often lacking that would allow the negotiation of what the local level can achieve, how it can be achieved, and what role the centre can be expected to play.

As outlined above, the Rural Tax and Fee Reform is an intriguing subject of study because, on the one hand, it was designed to somehow alter these structures and processes. On the other hand, however, it had to be implemented within these exact rules of the political game. This study argues that the reform failed because it only reduced local autonomy. Competition under hierarchy that had so severely affected poor rural governments, however, remained in place. The centre again provided reduction targets to achieve, but left it to the provincial and local levels to compete in working out the details. This put local governments into an awkward position: they were forced to compete in finding ways to reduce their own power and autonomy. While the central government probably sought to combine the positive aspects of hierarchical steering and competition to implement nationwide a sustainable policy sensible to local contexts, policy makers most likely underestimated the negative aspects of these instruments. The consequence was that some took up the challenge and came up with meaningful reform proposals, while others chose to resist because they either lacked the means to do so or had from the start not supported this policy. Eventually, resistance was so severe that the initial policy design had to be abandoned.

This outcome is not surprising and resonates with Heilmann's observation that experimental politics is not suitable for policies where local governments receive no tangible benefits. What is surprising, though, is that the central government attempted the reform at all. Or in the words of Linda Chelan Li, a well informed observer of the reform: 'the remedy for earlier failure of control was . . . yet more control, though it was not clear how it could possibly work better this time'.[64] Equally surprising is the fact that not all county- and township-level governments resisted the reform, which after all had originated in local experiments.

The argument

This book provides explanations for these puzzles by tracing the relationship between agency and institutional development in the RTFR from its very beginning up to its conclusion in 2006. It draws on an extensive evaluation of Chinese-language primary and secondary material, quantitative data, internal (*neibu*) publications as well as more than 100 interviews conducted with peasants, scholars, reform protagonists and cadres at all administrative levels (see appendix), and looks for answers by focusing on the county level. The county level is chosen because the counties had enjoyed considerable discretion in designing and implementing policies even before the reform, but were even further strengthened in the implementation phase. The understanding of how agency and institutions both internal and external to individual counties affect policy-making at this under-researched level of government is crucial for delineating the chances of central development programmes to be realized and own development policies to be initiated. Given their autonomy and relevance, the counties must in fact be considered China's primary local-level development agencies. The present analysis will provide insights not only into the county's role in policy-making, but also the nature and determinants of differences within counties and the endogenous and exogenous factors that hinder or further their ability to adequately play the role of a development agent. In particular, the centre's leverage over this important administrative level is examined, and their capacity to resist and even sabotage, but also to faithfully implement central policies is highlighted. Hence, on a more theoretical plane, it sheds light on the relationship between the centre and the localities as well as between state and society. In addition, it provides indicators for the central state's capacity to actively shape these relationships.

The following chapters scrutinize the factors responsible for the peasant burden problem, explain where and how the reform originated, analyse its design, shed light on its implementation and measure its outcomes. The book examines the interplay between (collective) actors, the division of rights and responsibilities, and the utilization of governing instruments in designing and implementing the reform. Thereby, the incentives facing political actors especially at the county level can be comprehensively outlined, and the centre's narrative that peasant burden and reform failure were solely the result of incompetence and a lack of moral fibre on the side of local-level officials can be challenged. In fact, the book will show how the central government skilfully obfuscated its own role in generating the burden problem and instead played peasants and local cadres out against each other.

As pointed out above, the nationwide implementation of the RTFR was implicitly justified as a necessary correction of the structure of central–local relations, which had become costly in monetary and social terms. Chapter 2 starts out by examining these costs. Specifically, it traces the evolution of bureaucratic sprawl and mounting local-level liabilities, which had been comprehended as being closely related to a high peasant burden. The chapter confirms that employees, fiscal outlays, and loans all increased dramatically after Reform and Opening. However, it also shows that the 'agency problem' can only be made partly responsible for these phenomena, and that the peasant burden concept is a political construct that obfuscates more than it reveals.

Chapter 3 further clarifies these points and at the same time lays the groundwork for the subsequent chapters by focusing on central–local relations before the RTFR. It illustrates that decentralization after Reform and Opening was not as complete as commonly assumed and examines the political, economic and social consequences of competition under hierarchy. In sum, Chapters 2 and 3 show that the so-called peasant burden was primarily a function of inequality in both local government revenue and peasant income coupled with an institutional set-up that fostered unbridled competition for resources and career advancement.

Against this backdrop, Chapter 4 looks into the agenda setting process for rural fiscal reforms to see which of these problems were addressed, and how. It distinguishes between two agendas, namely the peasant burden agenda and the rural fiscal reform agenda, and two arenas of discourse, namely the print media and the Party and government apparatus. It shows how the centre formed a vocal alliance with the peasant strata by shedding the blame for the 'burden problem' solely on local cadres. On the other hand, although local actors were substantially involved in bringing rural fiscal reforms to the centre's attention, this development was not at all mirrored in the media. It follows that these initially clandestine and dispersed bottom-up initiatives were aimed only at changing the structures of distinct localities, and no nationwide forum existed to coordinate such efforts and discuss the functionality of the overall institutional set-up.

Chapter 5 examines when, why and how the central government formulated a unified policy based on the evaluation of these dispersed experiments. Given that the conducting and evaluation of local-level experiments had been part of central-level agenda-setting, a variety of reform models already existed at this stage. The chapter introduces the three basic reform models and explains and evaluates the Rural Tax and Fee Reform model that was finally chosen. Regarded through the lens of this study's analytical framework, the initial reform design was clearly geared towards a significant recentralization of the Chinese polity. What is found absent, however, are attempts to change the instruments utilized by the centre to steer policy implementation.

Chapter 6 traces the implementation of this policy in Anhui province, where most of the experiments had originated, and beyond. Its particular focus lies on the local implementers, their attitude towards the RTFR, the leverage they were given and the instruments they employed in fine-tuning the policy to their local contexts. Reform implementation is being described as a 'tug-of-war' between the central government, who wanted the local levels to bear the costs of the reform, and resisters, who opined that this was the centre's responsibility. In between were the pioneers, who experimented with policies that, if proven successful in their particular county or township, were forced by the centre on all other localities as well. In the end, the resisters made the day, not the least because a majority of the peasants was not enthusiastic about the reform either.

Chapter 7 applies quantitative methods to measure the distributive effects of the reform on the peasant burden, peasant income and government finances at the aggregate county level in China's Anhui province. Thus, the results derived from the qualitative analysis of the previous chapters are substantiated by means of a statistical analysis. Specifically, the chapter explores and compares the parameters that determined burden reduction, peasant income and government finances during the initial phase of the RTFR, and after the agricultural tax had been scrapped. It is shown that, unsurprisingly, the wealthier counties were able to handle the impacts of the first reform stage much better, while the poorer counties had to deal with severe revenue losses.

Chapter 8 links the findings of the previous chapters and draws up a model of the reform process. The model illustrates how the failures and deficiencies in each reform stage were piled upon each other to gradually move the outcomes of the reform away from its initial intentions, resulting in a policy failure. It posits that China's rural problems can only be solved if the local agents are motivated to subscribe to the centre's aims and that they are more likely to do so if they can take part in policy formulation. Thus, the chapter argues for greater stakeholder inclusion not only in the agenda-setting, but also in the formulation of rural development policies.

2 Bureaucratic sprawl, government debt and the peasant burden: how serious?

As outlined in the previous chapter, the chief aim of the RTFR was to reduce fiscal exactions and thereby increase peasant income. It went further than previous attempts at burden reduction because it aimed not only at cutting taxes, but also sought to rid the political framework of the incentives for irregular fee-taking. What were these incentives? As Chapter 4 will show in detail, policy makers diagnosed a lack of accountability and transparency at the local level and found that central control over local governments was slipping. This, the narrative concluded, made it easy for greedy local cadres to take money from the peasants. Of course, most of this money was not pocketed for private gain, but was used to meet the ever-increasing expenditures of local governments. They were accused of having exhausted all other sources of revenue. Practically all county- and township-level governments were in the red, and many townships had accumulated liabilities so massive that they were practically bankrupt. Allegedly, unnecessary wage expenditures took by far the largest share in local government expenditures. As a consequence, policy makers at the centre advocated imposing checks on local governments and reducing their autonomy. The following sections examine the substance of these accusations. Was the increase of the local workforce, of wage expenditures and of local liabilities, really so massive? If so, is there a connection between excessive hiring and liabilities? And finally, is there really a relationship between these phenomena and increasing the peasant burden?

The cost of an expanding bureaucracy

How big is China's government workforce? First, we must distinguish between employees on the 'administrative payroll' (*xingzheng bianzhi*) and those on other payrolls, such as education and public services.[65] Only those on the administrative payroll can be considered 'cadres'[66] and enjoy the lifetime tenure and other benefits associated with this status. If only they are counted (as is the case in official statistics), China's government workforce is puny when compared to other countries. Apologetics of China's state sprawl thus frequently point out the fact that there is only one public servant per

150 citizens, which is far less than in most other countries, developing or non-developing.[67] However, if personnel on the other payrolls are figured in as well, China's citizens per public employees stands at 1:28,[68] or about 3.6 per cent of the population. Yet even this is small by OECD standards. Admittedly, such comparisons do not make much sense,[69] but these figures highlight one crucial characteristic of China's public sector employment: on average, the number of statistically recorded public sector personnel counts less than one fifth of the government's total workforce.

Bureaucratic expansion

Of more relevance to this study than the absolute size and the composition of the number of people 'eating the emperor's grain' is its increase over time. In this respect, the centre's allegations are confirmed. The number of public employees has indeed exploded since 1978. On the national level, the number of statistically recorded 'administrative' employees more than doubled between 1978 and 1993, but stagnated thereafter (see Figure 2.1) because a central government circular forbade the hiring of additional administrative employees.[70] As John Burns points out, however, the statistics hide a zero-sum shift in employment: while mass organizations massively shed personnel, the 'core government' workforce actually increased.[71] Nevertheless, the total number of civil servants remained nearly constant until 2002.

Wage expenses, however, have almost tripled during that same period, outpacing both inflation and the increase rate of government revenue. Note that the number of Party and government employees has been on the rise again since 2002, the year the Rural Tax and Fee Reform was implemented nationwide. In addition, wage outlays have increased almost 70 per cent

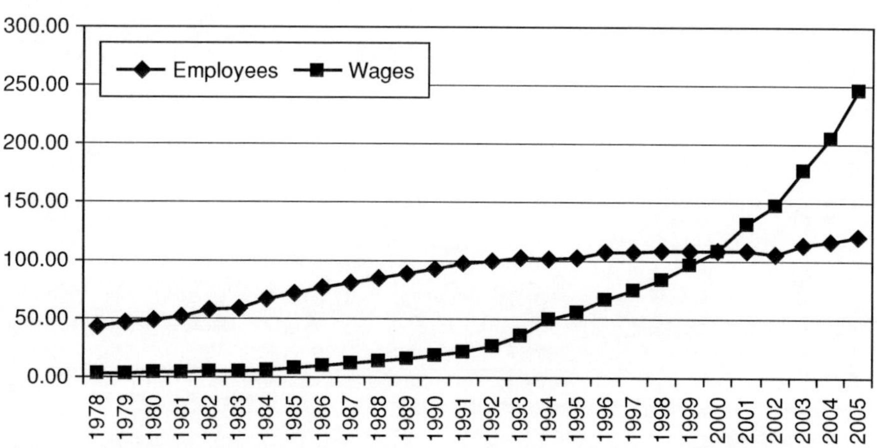

Figure 2.1 Staff in government, Party and social organizations, and their wages, 1978–2005

between 2002 and 2005. This is ironic given that one chief aim of the reform was to reduce administrative outlays by cutting personnel.

County and township employment was chiefly responsible for this steep increase. It is worthwhile to contrast the overall increase in civil servants (excluding service personnel!) with that at the rural level (township and below). While the former rose at an average annual rate of 4.3 per cent, the latter averaged 17.44 per cent. As Xie Qinkui shows, for example, the number of governmental units in Hainan's Wanning County more than tripled between 1972 and 1990.[72] In Hebei province, there was one government employee for sixty-two people during the Cultural Revolution (1966–76, all payrolls included). By 1995, one employee served forty-two people, and thirty people only eight years later.[73] In eight counties of Shanxi province, the ratio of citizens to public employed did not reach 1:20; in Daning County, the ratio was as low as 1:13.[74]

As seen in Figure 2.2, large-scale hiring commenced in 1983, as soon as China's administrative reorganization had started. That year, the communes were replaced by townships and the brigades by villages. As will be discussed in more detail below, this transfer entailed the restructuring of local government by redrawing the administrative boundaries and giving townships and villages more fiscal and administrative autonomy.[75] In addition, the management of the former brigade enterprises was handed over to the newly formed villages, and the townships became the new masters of the former commune enterprises.[76]

This problematic mix of newly gained freedom and largely intact centralist command structures led to an explosion of local government personnel which totally ate up the increase in local government revenue between 1997 and 1992. There were repeated attempts to cut down personnel and simplify local government structures, but they all failed.[77]

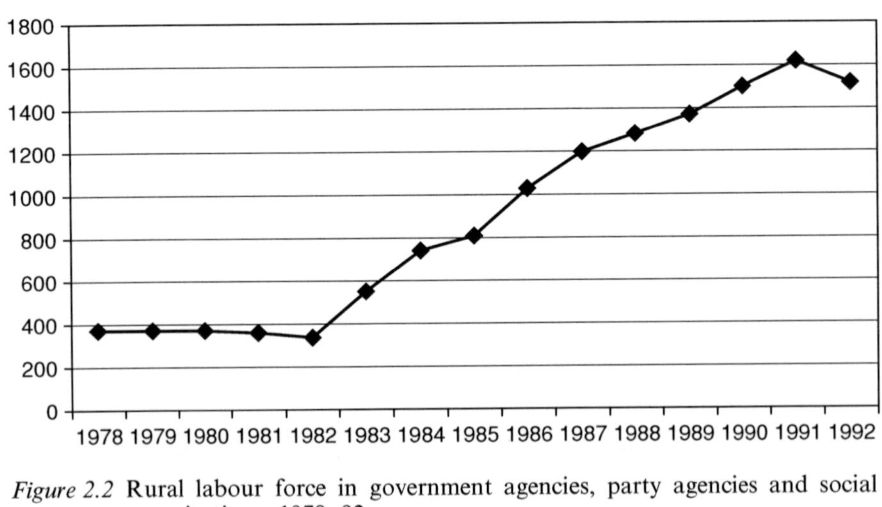

Figure 2.2 Rural labour force in government agencies, party agencies and social organizations, 1978–92

Attempts to downsize the workforce

The first attempt to reduce the number of persons paid out of government coffers fell on the years 1982 and 1983, and further reforms followed in 1988, 1993 and 1998. 'Each time, the local government workforce was reduced by 25 per cent to 50 per cent. The number of employees, however, quickly bounced back, ultimately exceeding its original number'.[78] In this quote, Zhiyong Lan obviously refers to the total amount of people eating 'the emperor's grain'. Figure 2.2 illustrates that rural administrative personnel was not affected at all by either the 1983 or the 1988 reform. Unfortunately, the Statistical Yearbooks after 1994 have stopped providing separate statistics on rural government employees, so we cannot assess the impact of the subsequent reforms for all of China. Accordingly, we need to make use of available case studies to assess post-1992 developments.

Comprehensive and verifiable case studies for the local level are precious and few, especially regarding the non-administrative workforce. Chen Xiwen, the Bureau Chief of the Central Leading Small Group on Rural Issues (*Zhongyang nongcun gongzuo lingdao xiaozu*), provides one such study. The material he presents on the personnel situation in Henan's Yanling County has the shortcoming of covering only the years after 1993, which makes comparison with the data just presented impossible. On the other hand, however, it describes not only the development of administrative employment, but also allows insights into the employment situation in the much larger public service sector, which includes teachers, the staff of agencies providing social services, blue-collar workers and hired labour (Figure 2.3). Employees of state-owned enterprises, however, are not included in these figures. They are on a separate payroll.[79]

Figure 2.3 illustrates many of the developments just described and allows further inductions. First, it confirms that in terms of workforce, the social service sector is much larger than the administrative sector. The service payroll was 9.4 times larger than the administrative payroll in 1994, but thirteen times larger in 2001! Second, in the administrative sector and, after 1999, in the service sector, more people had been hired than the payroll allowed. As will be discussed in the third part of this chapter, this 'excess' (*chaobian*) personnel put much stress on local government expenditures and, eventually, on the peasants. Note that before 1999, payroll positions exceeded the actual workforce. This is not due to underemployment, as government jobs are much coveted. Rather, in the absence of a social security programme for government employees, retirees were also put on the payroll.[80] Furthermore, for the purpose of fostering personal relationships, people who actually did not have to work were put on the payroll as well. According to Zhao Shukai,

> this is a common phenomenon. There are townships with up to 100 employees, but only twenty to thirty regularly show up for work. These people are on the payroll, they have a workplace, and they receive salary,

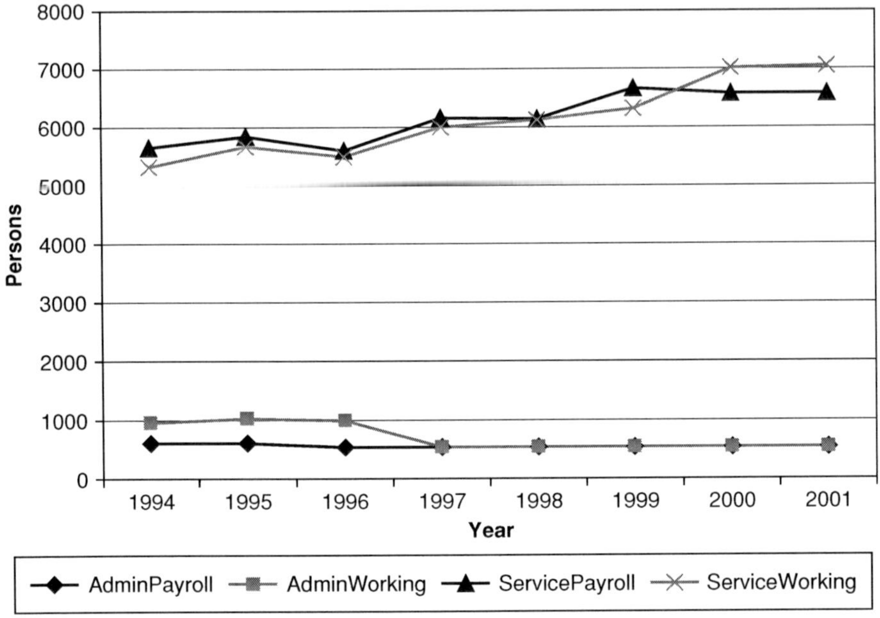

Figure 2.3 Persons on the government and service payrolls, Yanling County, Henan Province, 1994–2001

> but you can't find them. The township Party secretary and the township chief know that these people have nothing to do, but they don't know what they are actually doing. It's not that most of them don't show up for a long period of time. At any time, they might come to their office, walk around a bit and leave again.[81]

Third, while Yanling County in 1997 managed to get rid of 449 persons that exceeded the administrative payroll, 549 (!) persons were added to the service payroll. Obviously, superfluous employees were not let go, but transferred. Attempts to reduce the number of administrative units by merging townships, villages, and departments were successful, but scaling down the workforce was not. In Yanling County, for example, government bureaux were reduced from 490 in 1995 to 182 in 1997 and finally to 158 in 2000.[82] As Figure 2.3 makes clear, however, these reductions had no impact at all on the overall number of employees.

Wage expenditures and liabilities

Figure 2.1 gives an initial impression of the relationship between employment and wage expenses. After a period of rapid increase, hiring on the administrative payroll largely stagnated. Wage expenses, however, continued to grow at a fast rate. As a result, 7.8 per cent of total budgetary revenue in 2005 had to be spent on administrative payroll wages, up from 2.34 per cent in 1978.[83] These figures invite some number-juggling. Estimating that there are five times more employees in service than in administration (which is a low guess) and assuming equal wages between the payrolls (which is a high guess), we find that 60 per cent of all government budgetary revenue went into wages in 1994, when the service payroll was frozen. Due to the steep increase in government revenue, the ratio would have declined to 30 per cent in 2005. However, these numbers do not include teachers, whose salary eats up a large part of local expenditures.

In sum, it is not difficult to see that especially in poorer localities, the largest part of local budgetary revenue was spent on wages. In fact, there is a widely used and well known expression for the problem of local revenue being exhausted by personnel expenses. *Chi fan caizheng*, generating just enough revenue to pay government employees, seems to be a common phenomenon in poorer central and western China localities. Many do not even manage that, and wage arrears have become a common problem. Yanling County is a good example. 'In the whole county, there are 15,299 people [including teachers] who receive government wages. Each year, it needs 96 million yuan for wages alone, but its fiscal capacity (*shiji keyong caili*) lies at only 80 million yuan'. Accordingly, 'this county is unable to pay wages on time and in full. The problem of wage arrears (*tuoqian gongzi*) is serious'.[84]

An interview with the dean and the faculty of an Anhui model county village school revealed that the teachers had never received their full wages, and that wage arrears were a common problem. The school was so dilapidated that the author was not allowed to photograph it – the dean feared negative repercussions. The teachers were extremely unmotivated and complained not only about the heavy workload, but also the fact that some primary schools in the same county had received computer labs while they were still lacking basic necessities, qualified teachers included.[85]

All five townships and towns the author visited in northern and central Anhui suffered from high wage costs and oppressive debt. Central Anhui's M Town provides a good example. Revenue in 2004 added up to 3,580,000 yuan, and with 2,430,000 yuan, 'administrative costs' was the largest expenditure item. According to the head of the fiscal department, wage costs had also been hidden in other posts of the budget, such as the 'dissemination of documents' (1,190,000 yuan) or 'other expenditures (1,450,000 yuan).[86]

As a consequence, the township is severely indebted. The same official professed that the town's liabilities added up to 'more than 10 million yuan RMB, but we don't really know the exact amount'. This is more than three times the national average of 2.98 billion yuan for township-level liabilities,[87] and exceeded the town's revenue threefold. In addition, each village had on average incurred more than 200,000 yuan RMB in debts, which 'they will never be able to pay back' to the individual farming households that are the principal lenders. Despite this high level of indebtedness, the town can still obtain bank credits, even without any collateral. The key, the official stated, is the 'support' of the county, which can be easily obtained. Hence, he finds the 'debt situation in M Town not so severe' (*M. zhaiwu hai suan hao*). When asked what the borrowed money was used for, the official immediately named wage expenditures (*gongzi*) and administrative expenses (*bangongfei*). Apart from bank credits, M Town also generated funds by selling public property such as government-owned real estate, obsolete public buildings and schools that had been closed down.[88]

These observations indeed suggest that there is a direct relationship between a growing workforce and increasing liabilities especially in poorer regions. But are high wage costs the only source of local liabilities? Or are there other, maybe even more significant sources? The next section seeks to answer these questions by comprehensively examining the growth and composition of liabilities at the sub-provincial levels.

Growth and composition of liabilities

This section explores the relationship between local liabilities and the peasant burden. It starts by examining the debt level in today's China and then traces how, why and where local liabilities increased. Since national and local time-series data on the development of debt are not available, this chapter will have to rely on primary and secondary sources as well as interviews with veteran county- and township-level officials. It will be shown that there is a significant tendency for village administrations to lure peasants into lending them money with high interest rates. Frequently, however, such loans are not paid back, which indirectly increases the financial burden of the lenders. The determinants of how peasants are affected by high exactions are explored in the last section. Before we can devote ourselves to these issues, a short discussion of the data underlying this analysis is necessary. This not only helps to assess the robustness of the findings, but also provides a first glimpse into the administration of local governments.

The data problem

There are four basic problems with the available data on local liabilities. First, more case studies are available on central and western China than on the prosperous east coast. This is unfortunate, because the phenomenon of township- and village-level indebtedness is not confined to these parts of the country. The *Nongye Jingji Daokan* reported in 2002 that 84 per cent of China's townships were indebted, as were 81 per cent of all villages. On average, each of China's almost 50,000 township-level governments had to shoulder 4 million yuan of liabilities, each of the 740,000 villagers' committees 50,000 yuan.[89] Obviously, high revenue does not protect against liabilities. Zhu Gang, a researcher at the Chinese Academy of Social Sciences, has a simple explanation for this phenomenon: 'local governments in eastern

China have more revenue than those in western and in central China, but the upper levels also extract more funds from them'.[90] However, Zhu's recent and excellent research report on local-level debts does not contain any data on the more prosperous provinces.[91] Zhu justifies this by asserting that 'the situation there is not as acute. Also, it is much harder to obtain data there. Local governments don't cooperate. They fear that disclosing their debt level could hurt their market position, so they treat it as an economic secret'.[92]

Second, as is the case with the peasant burden discussed below, no universal definition of debt exists. Government statistics, for example, tend to equate debt with credits. Other forms of liabilities, such as unpaid wages, open contracts, the losses of state enterprise, personal loans and so on are usually not considered.[93] As will be seen, however, such liabilities make up a significant part of the total and should therefore not be neglected.

Third, there is no comprehensive data on the fiscal situation at the grassroots level, and the figures local leaders submit to the higher levels are chosen strategically.[94] If reaching certain fiscal targets is rewarded by central government subsidies, a poor township might suddenly prosper. For the allocation of poverty alleviation subsidies, however, the same township might lapse into poverty over night – at least on paper.[95]

The fourth point concerns the internal workings of local governments. Zhao Shukai, a researcher in the State Department's Development Research Institute (*Guowuyuan fazhan yanjiu jijinhui*), made first-hand experience as a county Party secretary in 2004. He states that local debt figures are based on estimates, and that reliable statistical data simply does not exist. Liabilities accumulated over a process of twenty years, and leaders come and go in quick succession. They are usually not too interested in the size of debt in their county or township because they do not stay long and there are no incentives for reducing debt. In addition, township-level cadres used to be recruited from the peasant strata in earlier years. They lacked accounting skills and did not understand how to calculate their township's liabilities. According to Zhao, this problem has not been solved. Township-level accounts tend to be sub-standard and sketchy to this day.[96]

Putting public debt into perspective

When does public debt become a problem? As the following paragraphs will make clear, it is not so much the absolute or even the relative amount of debt that matters, but rather its quality and its composition. For example, official figures put China's public debt in 2006 at 21.1 per cent of China's GDP,[97] far below the European Union's Maastricht Treaty security level of 60 per cent.[98] Nicholas Lardy doubts this figure and estimates public debt to equal 45–70 per cent of China's GDP.[99] In international comparison, however, even this higher estimate must be considered low. Public debt equals 64.7 per cent of the GDP in the United States, 98.3 per cent in Singapore and as much as 177.6 per cent in Japan, which ranks second highest in the world.[100] High public debt rates are not a problem if the lender is able to repay the loans, but they become problematic if he is not. Therefore, when evaluating the impact of public debt on public finance, not only the absolute amount of public debt and indicators such as the GDP/debt ratio, the debt per capita ratio and the individual income/debt ratio must be considered, but also the ratio of Non-Performing Loans (NPL).

China's local NPL ratio is very high. At the end of 2001, 43.97 per cent of all loans by the Agricultural Bank of China were non-performing. As a result of branch closures, massive firings and 'reducing its business in townships, consolidating its business in county seats and expanding its business in urban areas',[101] the NPL ratio decreased to 26.17 per cent in late 2005.[102] The chief credit source of China's peasants, the Rural Credit Cooperatives (RCCs), suffered from a NPL ratio of 37 per cent in 2002. In comparison, Japan's financial system was considered 'unhealthy' with a NPL ratio of only 8.4 per cent in 2002.[103] However, as will be explained in more detail below, loans from financial institutions form only half of all liabilities at the grassroots level. The other half is directly and indirectly borrowed from individuals. This goes to show that the debt situation in rural China is indeed very serious, and that the unhampered borrowing by local and grassroots governments contributes to increasing the financial burden some of China's farmers have to shoulder.

Debtors and creditors

According to government reports, the official figure for accumulated debts at the county and township level lies at around 1 trillion yuan RMB, which equals about 10 per cent of China's GDP in 2002. Where does this debt occur? An internal government publication states that 731 (35.6 per cent) of all counties are indebted,[104] but does not say to what extent. Given the absence of comprehensive data, this section will build on two very instructive surveys to examine the lenders and debtors, and to see how the borrowed money is used. One is a recently published report of Anhui Province's People's Congress Financial and Economic Committee. It is based on a survey of fifty counties and 125 township-level governments in seventeen cities of Anhui Province. With 66.7 per cent of all counties and 7.1 per cent of all township-level governments in the province investigated, this study rightfully claims representativeness.[105] The comparison with official data on average township and village debt in each of Anhui's prefecture-level governments in 1998[106] confirms the credibility of the survey.[107] In order to put the 'Anhui experience' into national perspective, the data are contrasted with an even more recent study of altogether eleven townships in central and western China conducted by Chinese Academy of Social Sciences researcher Zhu Gang. While the Anhui data can be considered representative, it is only so for one province, and their aggregate nature conceals the variation of the

Table 2.1 County, township and village share in total liabilities, Anhui Province, 2003

Administrative level	*Amount (billion yuan RMB)*	*% of total*
County	11.4	54
Township	6.5	31
Village	3.2	15
Total	21.1	100

Source: Zhongguo Shekeyuan Caimaosuo Ketizu, *Difang zhengfu zhaiwu: laiyuan, goucheng, yingxiang ji huajie duice* [Local government debt: origin, composition, impacts and counter-policies], Beijing: Chinese Academy of Social Sciences, 2004.

individual cases. Zhu Gang's study, however, covers central and western China and allows insights into each individual case, but due to the small number of cases, it cannot be considered representative. Nevertheless, the unique and detailed insight into altogether eleven townships and twenty villages in central and western China provides valuable clues about the variation that might be hidden in Anhui's aggregate data. Furthermore, it is of great heuristic value because the comparison of the individual cases allows the formulation of hypotheses regarding the formation and impact of local indebtedness. Table 2.1 lists the aggregate amount of debt incurred by Anhui's counties, townships and villages.

As can be seen, county-level liabilities make up more than half of all liabilities in Anhui province. This is an important finding. So far, most studies concerned with local government liabilities have focused on the township level, and not much is known about the situation in the counties. Quite possibly, the level of county debt has been underestimated. If we contrast this figure with the combined revenue generated by Anhui's counties in 2003, we find that liabilities exceed revenues by 58.3 per cent![108]

Zhu Gang's study, which unfortunately provides no county-level data, shows that we can indeed expect the variation within the sample to be significant, and that the situation in central China probably differs from that in western China. In his six central China cases, on average 86.3 per cent of all township-level and village debt is incurred at the township level, but the variation is as high as can be and ranges from 2.7 per cent to 97.1 per cent. In the western Chinese cases, the township level's share of the surveyed liabilities is 40.4 per cent, with a variation from 5.8 to 65.1 per cent.[109] As for the amount of debt, the situation is clearly more severe in central than in western China. The average sample central Chinese township is 37.82 million yuan in the red, almost thirteen times the national average given by the Ministry of Agriculture, and ten times that of Anhui Province. Again, the variation is huge, ranging from 352,000 yuan to 144 billion yuan! Average village debt stands at 545,000 yuan, roughly three times the national average and almost five times that of Anhui, but is also subject to immense variation. The debt levels in the western Chinese samples, in contrast, lie well below the national average. The respective figures are 1.8 million yuan (townships) and 31,000 yuan (villages), but the variance is equally large.[110]

Where did all the money come from? According to Zhou Xianbo, a researcher at the Anhui Academy of Social Sciences, there were sequences of local debt formation, with different lenders for each sequence. The first sequence corresponds to the surge in TVEs, which were mainly funded with loans from state banks and village financial cooperatives. When these avenues were closed, local governments turned to other entities for help, borrowing from state enterprises, private businesses and even foreign banks. Having exhausted these sources as well, village- and township cadres finally borrowed from individuals, often at very high interest of up to 20 per cent per year. The lenders generally were wealthy villagers and entrepreneurs and often even the cadres themselves. By signing their own loans at high interest, they were able to make a sizable profit. According to Zhou, township and village cadres and their relatives make up 70 per cent of all creditors, but unfortunately he fails to inform us about the origin of this figure.[111]

Last but not least, other sources of revenue were the various fees and fines levied from the villagers as well as IOUs on grain payments, defaulting on government-signed contracts, withholding wages and auctioning off government property. Zhou Xianbo observes a worrisome tendency towards the 'privatization' of state property: many farmers cannot pay their taxes and fees, so they lease their land to the creditors. The latter might refuse to pay the agricultural tax for their own land and that leased from the peasants on the grounds that the village or township also owes them money. This, however, prevents the grassroots administrations from meeting their tax quotas, so they apportion the loss among the villagers on a per capita basis, which further increases the peasant burden. When the land is finally concentrated in the hands of these 'new landlords' as a result of this vicious circle, they start buying schools and irrigation stations.[112]

Available case studies substantiate many of Zhou's propositions, but again some differentiation is in order. First of all, the debt phases were not as clear-cut as Zhou suggests, and villages and townships seem to borrow from different sources. According to the chief of the financial bureau in Anhui's M Town, village cadres more frequently borrow from individual persons than from banks or higher-level departments.[113] He conceded that it had become more difficult for M Town to obtain bank loans, but not impossible. The town government's main lenders are still banks and the county finance department.

The composition of the lenders of the accumulated debt of the villages, townships and counties of Anhui Province is given in Table 2.2. As can be seen, the bulk of the 21.1 billion yuan RMB in total liabilities is incurred with individuals. If one adds the categories 'credit from other entities and from individuals' and 'non-payment of grain subsidies', they add up to 67 per cent of the total. Loans from banks and other financial institutions

Table 2.2 Composition of county-level liabilities in Anhui Province, 2003

Item	*Amount (billion yuan RMB)*	*% of total*
Loans from superordinate departments	2.15	10.2
Open accounts from non-payment of grain subsidies	6.14	29.1
Credit from Chinese financial institutions	3.34	15.8
Credit from foreign financial institutions	0.74	3.5
Credit from other entities and from individuals	8.00	37.9
Other debts	0.73	3.5
Total	21.10	100.0

Source: Zhongguo Shekeyuan Caimaosuo Ketizu, *Difang zhengfu zhaiwu: laiyuan, goucheng, yingxiang ji huajie duice* [Local government debt: origin, composition, impacts and counter-policies], Beijing: Chinese Academy of Social Sciences, 2004.

represent only less than one fourth, and merely 10 per cent are borrowed from the prefecture, the province and the centre.

Zhu's study confirms that loans from individuals are an important credit source for the central Chinese villages, where it forms 40.19 per cent of all debt. Only two villages in the sample did not borrow from individuals. The situation is very different in western China, where the sources of village debt are so diversified that no common pattern is visible.[114]

In three of the central Chinese townships, loans from individuals made up less than one-fifth of total loans, while the other three did not borrow from individuals at all. The same applies to defaulting on construction contracts, which only three townships did at a significant scale (24–27 per cent of all loans), and two not at all. Only one town had accumulated significant wage arrears, which made up 16 per cent of the total. More important for all but one township in central China were loans from financial institutions, with a rate ranging from 16.25 per cent to 76.54 per cent of all loans. This loan category has been of importance to all villages as well, where it averages 22.9 per cent of all debt. Loans from higher levels were the second most important item, and each of the six townships had incurred such loans. Remarkably, neither the central nor the western Chinese villages did receive any loans from above, and if they did, they were negligible.[115]

In the western Chinese sample towns, the situation looks quite different. Each one of them had defaulted on construction contracts. This made up half or more of the debt total in three cases, and between 11 per cent and 26 per cent in the other two. In one town, debt consisted mainly of wage arrears, which approached 50 per cent of all debt, but this item was comparatively unimportant in the other four cases. Loans from financial institutions and loans from higher levels were equally unimportant.[116]

Different as theses samples are, they confirm that borrowing from individuals is the most important source of liabilities in central China's villages.

In addition, they imply that loans from financial institutions and fiscal departments are important for central China's townships, whereas western Chinese townships tend to incur debt by defaulting on construction contracts.

While this is not the place to muse about the differences just observed, it is important to note that uniformity with respect to either the level of debt or the composition of the creditor and debtor groups should not be expected even within one province. This large variance testifies to the next section's observation that inequality is an integral part of the various 'problems' researched here. Zhu Gang's data allow another relevant observation: as a general rule, the peasant burden seems to be more serious in places with high debt levels. Simple correlation analysis confirms this: the Pearson correlation coefficient for debt and burden ratio is $r = 0.96$ for central and $r = 0.98$ for western China. Accordingly, the values of r^2 are very high for both central (0.94) and western China (0.96), meaning that more than 90 per cent of the variation in the peasant burden can be explained by the variation in local debt.[117] In other words, Zhu Gang's data provide support for the hypothesis that there is not only a causal relationship between high debt levels and the peasant burden, but that the debt level *determines* the burden level![118]

Reasons for indebtedness

In order to close the circle and link the local debt situation to the institutional incentives for overspending as well as to an increased peasant burden, the following paragraphs will elaborate when the debts just described were incurred, and what they were used for. Zhao Shukai, who was introduced above, had first-hand experience as a county Party secretary in 2004. According to Zhao, who painstakingly calculated the debts his county had incurred, there were three major stages of debt formation.

The first stage commenced in the mid-1985s and was intimately connected to the aim of developing the countryside by encouraging decentralized industrial development. Township and village enterprises (TVEs) mushroomed, but were largely financed by bank credits, as the next chapter will show. The second stage began in the early 1990s, when the central government laid much emphasis on the development of the rural infrastructure, which it tried to achieve by means of competition under hierarchy: the evaluation of local cadres was linked to the fulfilment of targets set by the centre (*dabiao*). School-building, compulsory education, road construction and investments in agriculture were all subject to such targets. The third stage commenced in the late 1990s and lasted until the beginning of the RTFR a few years later. Debt at this stage, according to Zhao, was incurred because of excessive government consumption. This included the investment in projects designed to attract business, and expenditures for lavish banquets were not a small item in this category either. Whereas banks and rural financial institutions had still financed the expenses of the first two stages, they were not able to do so in the third stage. Instead, local governments delayed or stopped the paying of wages and started to borrow from individuals.[119]

The comparison of two towns in G County is very instructive, as it shows the long-term impact that TVE development, the hiring of personnel, and target achievement programmes had on the development of local debt.[120] X Town is a bit larger in size than C Town, and the population in both towns is predominantly rural. The biggest difference is that X Town had invested significantly in TVEs, whereas C Town had not. Accordingly, X Town had a good head start on incurring debt. In 1996, its liabilities had already added up to 3.4 million yuan, 70 per cent of which resulted from investments in TVEs. At that time, C Town's liabilities stood at only 159,000 yuan, most of which had been used for the provision of basic public infrastructure.

Less than ten years later, the picture looked dramatically different. In 2004, X Town had incurred a total of 35.5 million yuan in liabilities, 11 million of which was in accumulated interest alone. The larger part of this interest stemmed from loans to finance the development of TVEs. Principal and interest combined, this item accounted for almost half of total liabilities. Funds borrowed for wage payments followed closely, and project expenditures such as for education and birth control ranked third. 'Hidden debts' such as unpaid IOUs and construction costs, and 'everyday expenditures' were the smallest items.

Due to the absence of TVEs, C Town's debt composition is very different from that of X Town, and the debt level is much lower. Principal and interest combined, C Town's debt has increased from 159,000 yuan in 1996 to 18.7 million yuan in 2004. More than half of all liabilities were incurred in the process of refinancing the RCCs, a rather small item in X Town. The second biggest item was project expenditures, densely followed by loans for wage payments and for buying office equipment.

Apart from the differences in these accounts, there are two remarkable similarities: first, local liabilities did not increase steadily, but exploded after 1997. Not incidentally, this happened just after the central government had tightened the squeeze on extra-budgetary funds (see next chapter), while simultaneously demanding an increase in project accomplishments. According to the head of a northern Anhui township's Bureau of Finance,

> the year 1997 was a turning point for fiscal issues. By 1999, we were unable to pay wages regularly, and wage arrears started. The introduction of the tax-division system was only one reason. The key issue was that in 1997 and 1998, 'assessment and comparison' campaigns were conducted on a major scale, such as target-achievement programmes for urbanization measures, the 'two columns' of education, raising the cultural level, infrastructure building, and so on.[121]

Second, funds borrowed for unpaid wages and construction projects figured very prominently in both towns. However, even this small sample of cases cautions us to not engage in simplistic generalizations. For example, although rural governments frequently default on wage payments, it has been shown above that loans incurred to address this problem do not necessarily fare big on the township debt tab. Zhu Gang's samples from central China did not borrow to pay wages; neither did the majority of the western Chinese cases. Furthermore, as the cases just described also show, there is no clear relationship between TVE related debts and overall liabilities. The assumption that large loans for the development of TVEs or RCCs correlate with high overall debt levels is refuted, as is the hypothesis that such loans correlate with a high level of accumulated interest (presuming that they were solicited in the first debt phase just discussed, more than twenty years ago). As the meagre data available suggest, however, there seems to be one independent variable with powerful explanatory value for local-level debt: central-level policy changes, especially where they provide incentives to invest or spend. The next chapter will continue to provide evidence that policies by the central government have significantly contributed to the problem of over-taxation. Before that, it is necessary to take a closer look at a concept that is frequently mentioned, but ill defined: the peasant burden problem.

The peasant burden problem

Since the mid-1980s, the 'peasant burden problem' (*nongmin fudan wenti*) has been prominent on the political agenda and there is a widespread agreement that the problem has worsened. No doubt, the previous sections seem to lend credibility to this assumption. However, a closer look shows that the peasant burden problem concept is in itself problematic and needs differentiation, especially when it is to be applied in analytical research. The following paragraphs will elaborate on the concept and its empirical foundations. Specifically, they will show that:

1 Data are lacking to support the claim of a 'burden problem' in rural China.
2 No agreement exists on the conceptualization of 'burden'.
3 No comprehensive evidence is presented to prove that burden has indeed increased.
4 'Burden' is not only a function of local government exactions, but also one of income inequality. Relatedly, the problem does not stem so much from excessive as from unjust exactions.

Concept

How serious a problem is the peasant burden? Given the great consensus that exists in academia on the stability-threatening nature of the issue, data are remarkably scarce. Even the excellent, data-rich monograph on the peasant burden by Thomas Bernstein and Xiaobo Lü finds that '[i]t is difficult, if not impossible, to come up with a simple number' of 'how much . . . the average peasant household had to pay at any one time'.[122] Unfortunately, comprehensive and micro-level data on peasant income, fiscal and non-fiscal exactions, the inflation rate etc. are often not available, and if they are, they are not exact. First, it is a common practice for officials at all levels to cook statistics to meet targets set by higher governments, to feign underperformance in order to be eligible for subsidies, or to under-report revenue and pocket the difference to sustain government operation or simply for themselves.[123] Second, the methodology and even the sample size of existing surveys are frequently not explained, which renders comparison between existing data sets risky. Finally, as is the case with research on local debt, most investigations on the excessive peasant burden and associated problems were conducted in central China. Studies on western China are much fewer in nature, and despite extensive archival work, the author has not been able to obtain any material on prosperous eastern China. Chinese specialists in these fields claim that such studies are not conducted, because 'the problem is not as acute there'.[124]

What does the peasant burden connote? A survey of the literature reveals a considerable lack of agreement. Some scholars employ the term to encompass selected government exactions, others include rising living costs as well. For example, increased medical costs or higher prices for farming inputs like water, electricity or fertilizer are often characterized as also contributing to the peasant burden. Although there is some justification in applying such a wide concept, it has proven difficult to operationalize. Rising prices do not per se constitute a burden, even in the case of basic necessities. The question is whether people can or cannot afford to pay higher prices. If the relationship between higher prices and income differences is not heeded, the concept becomes nondescript. Furthermore, even in the case of the narrower concept are definitions often not the result of theoretical deliberations, but of limitations in data. For example, most of the studies on this issue only look at singular features of the peasant burden, such as the rural taxes and quasi-taxes, but do not cover illegal exactions.

Four different kinds of government exactions must be differentiated. According to folk wisdom, 'the first tax is light, the second heavy, and the third and the fourth are a bottomless pit' (*yishui qing/ershui zhong, sanshui sishui wudidong*). 'First tax' stands for the real taxes levied, namely the agricultural tax, the tax on agricultural special products, the slaughter tax and the animal husbandry tax. The pre-reform agricultural tax rate was 15 per cent of the average harvest in the three years after the 'Agricultural Tax Regulations of the People's Republic of China' (*Zhonghua Renmin Gongheguo nongye shui tiaoli*) came into effect in 1958. As the rate was not adjusted to the rising yields of the subsequent years, in the 1980s it amounted to only about 3 per cent of the actual harvest.[125] 'Second tax' stands for the village retention (*cun tiliu*) and the township comprehensive fee (*xiang tongchou*). These important quasi-taxes were, in their combination (*san ti*

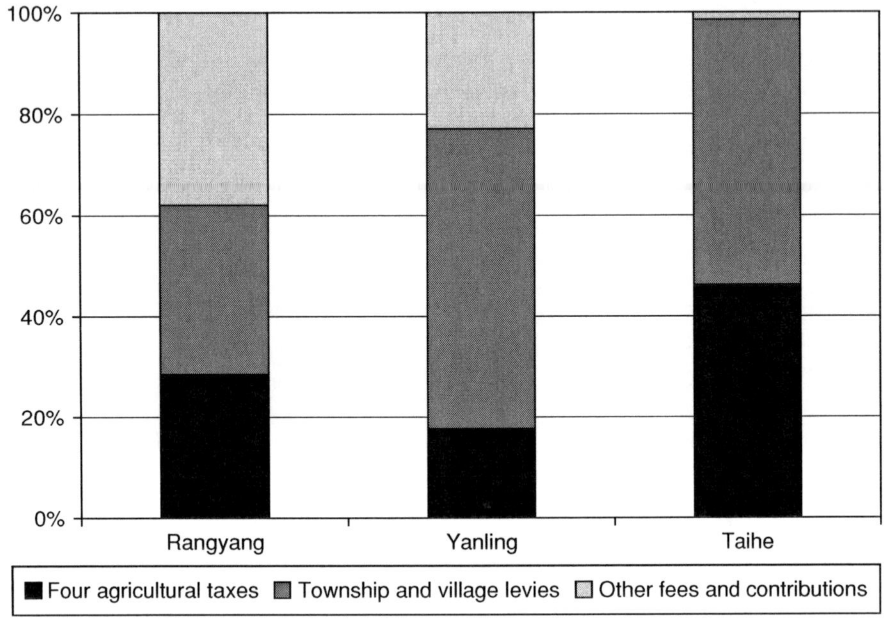

Figure 2.4 Burden composition in three central Chinese counties, 1999

wu tong),[126] not allowed to exceed 5 per cent of peasant revenue. The next chapter discusses these items further. Finally, the 'third tax' denotes irregular, but legal exactions such as the road construction fee, the education fund etc., and the 'fourth tax' stands for illegal fee taking. Naturally, the relative sizes of the 'third' and 'fourth' taxes are subject to large variations as well. A comparison of the aggregate burden composition (excluding illegal fees) in three counties illustrates this point well (Figure 2.4).

These figures confirm that except for Anhui's Taihe County, which had already been experimenting with a fiscal reform at the time of the survey (see Chapter 4), the 'first tax' was light, whereas the second was heavy. However, they seem to contradict the notion that the 'third tax' is a 'bottomless pit'. Nevertheless, it is the largest revenue item in Rangyang (38 per cent of the total), and exceeds agricultural tax revenue in Yanling. Since the 'fourth tax' is not represented in this graph, there is probably some truth in the peasant proverb.

Development

As for the development of the peasant burden, official statistics do not support the central government's narrative of a dangerous increase. On the contrary, they document a decrease in the peasant burden (defined as the sum of the agricultural taxes,[127] the *san ti wu tong*, 'societal burdens' (*shehui fudan*) and compulsory labour). In 1995, it averaged 12.5 per cent of peasant income, and steadily decreased in the subsequent years (to 11.1 per cent in 1996, to 10.3 per cent in 1997 and to 10.1 per cent in 1998).[128] According to the *China Statistical Yearbook* and the *China Agricultural Yearbook*, the peasant burden has steadily decreased from 12.2 per cent of peasant income in 1994 to 7.2 per cent in 1999.[129]

The picture is not much different with regard to scholarly contributions. An unnamed contributor to a recent and prominent publication on rural politics edited by the Chinese researchers Xiao Tangbiao and Li Changqin claims that the peasant burden has exploded since the 1980s. As evidence, he shows that the *san ti wu tong* have increased 13.8 per cent annually between 1993 and 1998, from a total of 44.6 yuan RMB per capita to 84 yuan RMB. With less than 5 per cent per annum, however, peasant income had increased much slower. This indeed sounds shocking, but this huge difference largely results form the fact that the income was cleared of inflation, but the fees were not.[130] If the nominal average peasant income for the respective years is used, we find that the relative burden from these fees has even decreased by almost 1 per cent.[131] Another monograph on the peasant burden lists a village's equally impressive per capita burden figures for the years 1981 to 2000, unfortunately without providing peasant income figures. Again, if the burden totals provided (consisting of the sum of the agricultural taxes and the *san ti wu tong*) are related to average peasant income during those same years, the burden increase is almost negligible (0.74 per cent between 1981 and 2000).[132]

A fine monograph on local finance and peasant income edited by the bureau chief of the Central Leading Small Group on Rural Issues (*Zhongyang nongcun gongzuo lingdao xiaozu*), Chen Xiwen, allows further insight into the 'burden problem'. According to official statistics, the average burden level in rural Huopai Town, which belongs to Rangyang County in rural Hubei province, decreased from 8.25 per cent in 1991 to 5.43 per cent in 2000.[133] On the other hand, a survey of fifty-five persons in that same county revealed an average burden level of 28 per cent, as opposed to 8 per cent in Anhui's Taihe County and 9 per cent in Henan's Yanling county.[134] Assuming that they are true, these figures lead to the following conclusions: first, the burden level seems to vary strongly from region to region[135] and most probably also from family to family. Second, and relatedly, both the high burden level discovered in the household survey and the large differences between the surveyed counties suggest that the problem is rooted not only in excessive exactions, but also in high levels of income inequality. This is confirmed by national figures as well as case studies. For example, official statistics confirm the regressive nature of the peasant burden (Table 2.3).

Clearly, poor peasants had to submit a far higher percentage of their income to the government than richer peasants. For the former, the burden was indeed a problem, whereas it was not for the latter. This is closely related to the income structure: the study also shows that peasants who only worked in

Table 2.3 Peasant burden for different income levels, China, 2002

Income/year (yuan RMB)	*Absolute burden (yuan RMB)*	*Burden as part of income (%)*
Less than 500	117.9	37.3
500–1,000	115.6	14.9
1,000–1,500	129.9	10.4
1,500–2,000	137.5	7.9
2,000–2,500	138.5	6.2
2,500–3,000	156.0	5.7
3,000–3,500	266.1	5.2
Over 3,000	254.9	4.9

Source: General Bureau of National Taxation, reproduced from X. Chen, *Zhongguo xian xiang caizheng yu nongmin zengshou wenti yanjiu*, Taiyuan: Shanxi jingji chubanshe, 2003, p. 120.

agricultural production had to sacrifice a significantly larger part of their income than those with off-farm incomes.[136]

Conclusion

Limited as they are, these data hint at some important tendencies. Similar to the level and sources of rural government debt, the peasant burden displays strong regional and individual variation. Interviews with peasants conducted by the author in Anhui and Hebei provinces have revealed that the inequality thus created is an important part of the problem, because the most affected groups feel marginalized and subject to unfair treatment.[137] In total, the peasants are much better off today than during the reign of Mao Zedong, and the average peasant burden is also lower. Under Mao, the peasant burden ranged from 20 per cent of peasant income in 1962 to 35.2 per cent in 1970.[138] Since there was little variation in personal affectedness, however, peasants tended to accept the burden as something everyone had to shoulder. Burden was heavy, but not considered a problem. The feeling of marginalization after Reform and Opening was amplified by the fact that poor farmers without sideline occupations were most disadvantaged by the fiscal system.

Therefore, the 'problem' was rooted not so much in the overall burden level than in the regressive nature of government exactions, which strongly disadvantaged peasant households with low incomes. Hence, the 'burden problem' clearly was function of income inequality. Peasant income inequality, however, was not primarily caused (though surely reinforced) by excessive exactions, but, as Chapter 7 will show, determined by the local economic structure and the absence of viable development policies. In sum, the section has shown that China's peasants are by no means a homogeneous group, and that not all peasants were equally affected by excessive exactions. In clear terms, it was the poor peasants who suffered the most.

Even if it is not possible to prove that the peasant burden has steadily increased, or that it has increased at all, there is considerable evidence that the phenomenon has indeed been conceived as a growing problem not only by the peasants, but also the general public. This issue will be thoroughly discussed Chapter 4, so suffice it to say here that the central government's subtle propaganda played a significant part in shaping this perception. The perception that a problem exists adds to that very problem, even if the person who claims that there is a problem is not affected by it. Prominent media coverage of impoverished peasants, burden-related protests and cruel exaction methods helped shaping perceptions of an unbearable peasant burden threatening social stability.

The analysis of government sprawl and local debt confirms the centre's narrative that both phenomena have worsened rapidly after Reform and Opening and that they indeed propelled irregular and excessive exactions from China's peasants. On the other hand, the analysis has also shown that personal greed and clientelist politics at the local level were only partially responsible for these developments. Especially in poorer regions, competition under hierarchy also bears a large responsibility for overspending and debt formation at the local level. The next chapter explores this issue further by looking at the structural causes of these developments.

3 The institutional sources of the peasant burden

This chapter examines the institutional sources of the government sprawl, local debt and the peasant burden that were described in the previous chapter and thereby lays the groundwork for the subsequent chapters. It enables us to contrast the results of the broad scientific enquiry of the phenomenon conducted in this and the previous chapter with the highly subjective political viewpoints and strategies championed by political actors at all levels during the whole reform process described thereafter, this allows us to enquire into the respective roles of institutional constraints and actor-specific strategies in the agenda-setting, formation, implementation and evaluation of the RTFR. With respect to agenda-setting, for example, it will be interesting to see which aspects of the 'burden problem' elicited in these chapters were highlighted for what reasons, and which aspects were not addressed at all. The same applies to the other stages of the policy process. Understanding the systemic sources of the problem enables the formulation of hypotheses about why certain deficiencies were remedied, but others not, and it enables us to better understand the role of actors and institutions in the implementation process. Against the background of this prior analysis, finally, the policy can be evaluated, and this scholarly evaluation can be contrasted with the centre's official assessment of the reform's impacts and outcomes.

The following sections examine in what ways central–local relations have changed since Reform and Opening, and what impact this had on the peasant burden. Attention will be paid to the structure as well as to the process dimension of the concept. As for the structural dimension, it is necessary to disaggregate the decentralization concept. Existing concepts usually differentiate between the sub-types of administrative, fiscal, and political decentralization.

Administrative decentralization is

> the transfer of responsibility for the planning, financing and management of certain public functions from the central government and its agencies to field units of government agencies, subordinate units or levels of government, semi-autonomous public authorities or corporations, or area-wide, regional or functional authorities'.[139]

Given that not only the responsibility to plan and manage, but also to finance certain public functions is sometimes devolved in the case of administrative decentralization, it is necessary to ask where the resources come from. This issue is addressed in the subtype of fiscal decentralization, which refers to the devolution of fiscal resource generation to sub-national levels of government.[140] Most important, it encompasses the regulation of central and non-central taxation, such as regional and local rights to levy taxes, the right to collect central taxes by sub-national governments for their own use, or to deduct additional regional or local rates on central taxes.[141] Apart from that, it entails the rights to levy user charges or fees, and to obtain bank or other loans.[142] Since the present study is concerned with issues of rural finance, this dimension is of crucial importance. Political decentralization

> refers to situations where political power and authority has been transferred to sub-national levels of government. The most obvious manifestation of this type of decentralization are elected and empowered sub-national forms of government ranging from village councils to state-level bodies.[143]

Aiming to give 'citizens and their elected representatives more power in public decision-making',[144] political decentralization can also be understood as the transfer of 'law-making power'.[145] Hence, the 'replacement of appointed officials with elected local or regional officials' generally is a defining feature of the concept.[146]

By examining each of these dimensions, the following sections will confirm our guiding hypothesis that the structure of central–local relations has changed significantly since Reform and Opening was initiated in the late 1970s, while the instruments applied to implement unpopular policies have not. It will be shown that conscious decentralization and unintended loss of control have combined with the persistence of competition under hierarchy to not only open up avenues for excessive exactions, but also to provide the incentives to exploit the peasants. It must be pointed out, however, that this especially applies to regions with a large agricultural base and little or no industry.

Functional division of labour

As for the functional division of labour between China's spatial administrations, it can be safely stated that the county and township governments have gained considerable autonomy after Reform and Opening, but not as much as is often assumed. Naturally, the level and scope of administrative decentralization depends on the kind of responsibility. Defence, for example, is a highly centralized matter, and the regional military commanders can be regarded as field agents of the central command with only a tiny scope of decision-making power.[147] With regard to rights and responsibilities more intimately related to the peasant burden, the situation is quite different. The following sections examine the three administrative fields identified as sources of the peasant burden, namely economic management, the provision of social services and the scope of authority to recruit, evaluate and promote personnel. Furthermore, as administrative central–local relations are inextricably linked to the modes of policy implementation, the latter will also be subject of this part's analysis.

Delegation of economic management

As stated in the introduction, the delegation of economic management to China's provincial and local governments was without doubt the primary factor for setting in motion China's sustained economic growth.[148] On the other hand, however, it had severe consequences for those local governments who were disadvantaged in the competition for investments and outputs. The following sections will elaborate on this observation in more detail. Since it would be beyond the scope of this treatise to summarize all the excellent work on the role of local governments in decentralized rural economic development, this section will restrict its focus on their role in promoting industrial growth. In particular, the developmental impact of community-owned rural Township and Village Enterprises (TVEs) is of particular significance for the present study.

After return to family farming in the late 1970s, which resulted in a remarkable increase in agricultural productivity,[149] the early 1980s were devoted to developing rural small-scale industry.[150] As increased marketization was desired, but privatization was not, the centre decided that the rural enterprises were to be run by the township-level and village administrations. Without a doubt, many local governments were able to use this transfer of property rights to their advantage, and the mid-1980s witnessed a mushrooming of TVEs all over rural China (Figure 3.1).

Figure 3.1 shows that the number of TVEs has steadily increased from 1983 on. That year, they started operating under the management of the villages and townships, which had themselves been put under decentralized management after their transformation from production brigades and communes.[151] After the Tiananmen Square incident in 1989, however, the growth of TVEs stagnated. As the popular dissatisfaction that had discharged in the mass demonstration in the heart of Beijing was partly attributed to double-digit inflation of the preceding years, the centre sought to improve the control of economic growth. As a result, features of the planned economy were revitalized, and a ban on new investments imposed.[152] After Deng had courted support for his liberalization programme with provincial party, government and military elites in 1992, growth resumed.[153] Because their turnover soon exceeded that of the state-owned enterprises (SOEs) in some provinces, the TVEs were hailed as transformers of the Chinese economy.[154] In addition, as Figure 3.1 above shows, they provided an important channel

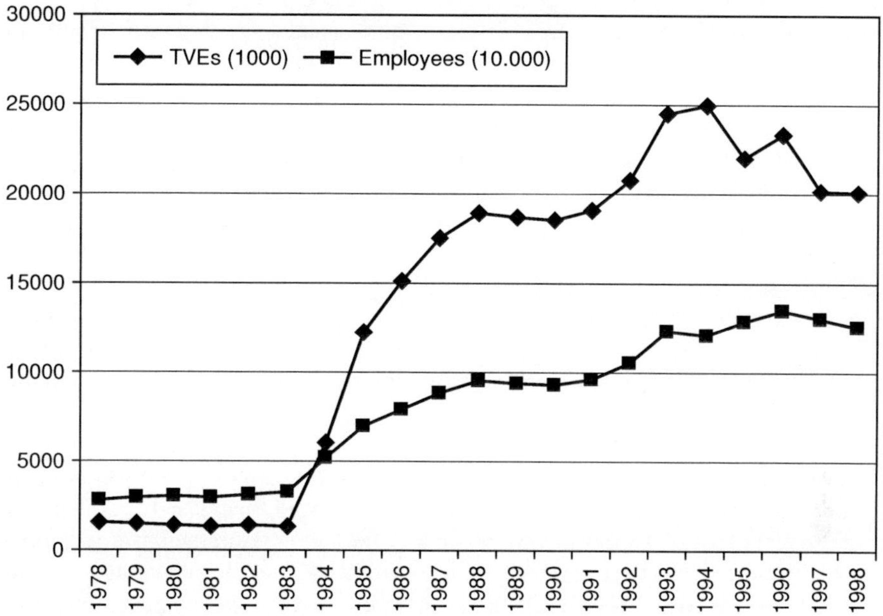

Figure 3.1 Number of TVEs and people employed by TVEs, 1978–98

for surplus rural labour. In 1996, TVEs employed 135 million people, more than one fourth of the rural workforce.[155]

As for the down side of administrative delegation in economic policy-making, it has to be stressed that delegation is not only about rights, but also about responsibilities. This is often overlooked in the discussion of China's economic take-off, and is one of the core reasons for the problems many TVEs soon had to face. Specifically, in striving towards realizing a 'planned commodity economy', the central government attempted to steer rural industrial development by setting production targets for local governments. Similar to the dynamics of the Great Leap Forward,[156] township- and village-level leaders soon found themselves in fierce competition over meeting and exceeding industrial production targets.[157] Naturally, after they had been exceeded, the targets were further increased. According to one official cited by Susan Whiting, targets in one county were raised 30 per cent in 1992, but 'old enterprises [were] doing well if they increase[d] 10 per cent a year'.[158]

Despite these pressures, TVEs brought many advantages to local leaders, but not necessarily to local governance. Not only did they initially boost the local GDP, the most important criterion of cadre evaluation, but they also created off-budget revenue which the higher level had no control over (see below). In fact, they were the single most important revenue generator for townships, which were allocated only a marginal share of local taxes. This had severe consequences for the incentive structures provided to local cadres. As many TVEs merely served as laundering machines for local revenue, their managers were not much concerned with demand structures or even profitability. According to Christine Wong, 'local governments try to maximize net revenues, which consist of profits and taxes paid by the firm. They would be willing to allow the survival of money-losing firms as long as sufficient tax revenues were generated to offset the losses.'[159]

In order to meet the targets specified by the higher levels, new enterprises had to be built, and they were paid for by loans from the local branches of the state-owned banks. Even where townships or villages could offer no or only insufficient collateral, the local branches of the state banks were, via Party channels, forced by the county or township governments to extend loans.[160] Obviously, as the Party secretary of a poor northern Anhui township confided, TVE debts were never repaid, no matter how profitable the enterprises were:

> At that time [the latter half of the 1980s], the forceful development of TVEs was promoted. Much debt was caused because the township government stepped in as guarantor for the TVEs. Back then, it was impossible not to do that. The upper level ordered them to guarantee for the TVEs, because if the township government did not act as guarantor, the TVE could not get any credit. In these years, the TVEs did indeed develop, but now they are basically all bankrupt. But actually, it does not matter if a TVE prospers or not, because the debt is never repaid anyway.[161]

In 1989, TVEs had accumulated 86.52 billion yuan in loan debts from the Agricultural Development Bank of China and RCCs, and the figure rose to 305.83 billion yuan in 1994, the last year for which statistics are available.[162] While the ratio of bank loans to fixed assets of China's TVEs stood at only 10 per cent in 1978, it had increased to 60 per cent in 1993.[163]

Delegation of social service provision

While throughout the 1980s, the central government's focus was mainly on rural industrialization, the 1990s saw a change in priorities to the improvement of the local social infrastructure. The year 1992, when a costly regulation demanding the provision of nine year compulsory education to all children was passed, marked a watershed in these developments.[164] Local governments became saddled with formidable responsibilities in the area of welfare. As will be seen, local social service provision is another policy field where the drawbacks of competition under hierarchy induced by administrative delegation by the centre become clearly visible. Here, the impact is especially heavy, given that, as a World Bank report points out,

> . . . the county is the single most important level in all the provinces, accounting for one-third or more of sub-national expenditures. . . .

> [C]ounties and townships account for a majority of expenditures in all the key items affecting the quality of rural life: education, health, public administration, support to agriculture, civil service pensions, and aid to poor regions.[165]

Together, they are responsible for almost 90 per cent of all expenditures for social services such as education, health, culture and science, and have to bear 99 per cent of all welfare expenditures.[166] Most of these services were, prior to the RTFR, provided and financed at the county and township (*xiangzhen*) levels. For example, these two levels bore around 70 per cent of all expenditures for education and 55–60 per cent of those for health.[167]

How was competition under hierarchy organized in this domain? First, no detailed regulations existed on who had to provide what services, and how. Similar to the guidance of rural industrial development just discussed, the centre formulated target figures (*zhibiao*) for social services which were deemed necessary. Already in 1993, at least forty-three target-setting programmes had been in existence.[168] These targets were then passed down by having local officials sign a number of 'position responsibility contracts' (*gangwei zeren zhuang*) that specified the targets to be attained.[169] The 'two columns' of education programme, for example, demanded the following requirements to be fulfilled for each village:

> Setting up a primary school in a two-storey building, replacing mud walls with brick walls and brick and wood constructions with ready-made concrete parts, cover the walls with porcelain tiles, convert the school yard into a garden, provide up-to-standard equipment.[170]

As during the Great Leap Forward, avid local cadres sought to exceed even these lofty standards. The headmaster of a primary school in a poor Shandong township, for example, proudly showed the author around a luscious garden. Upon entering the school, the visitor beheld a lovely 4 m replica of holy Mount Tai sporting miniature pagodas and a little waterfall.[171] As Bernstein and Lü state, '. . . the political reputation of counties hinged on designations as 'advanced', which prompted fierce competition among them'.[172]

Generally, the centre did not busy itself with the methods local governments used to achieve, or more often pretended to achieve, these targets. Of course, if the methods employed caused major local disturbances or even international protests, as in the case of violence related to birth control, the centre would intervene. However, it would only formulate another target demanding the cessation of, say, violence related to tax collection, birth control or the assertion of villagers' rights, while at the same time leaving the incentives for the use of violence intact. As one former county chief told the author, counties in Hebei Province (and probably in other provinces as well) prepare birth control rankings of all townships and towns for internal distribution. To be ranked at the end of the list does not only mean a loss of face, but often even demotion.[173] Rural cadres find this highly unfair. One agitated township Party secretary complained:

> How am I supposed to reach the [birth control] target if I cannot punish people? If I deal with violations according to the law, they will get a fine, but cannot pay it. So what is the risk for them to ignore the regulations?[174]

Personnel recruitment rules

To enhance their ability to engage in local economic development and fulfil their political tasks, hiring personnel above the quotas set by the centre had been tolerated at the beginning of the reform period, leading to a hiring spree by county- and township-level governments. According to Thomas Heberer and Wolfgang Taubmann, the average number of town cadres (not including other employees) in Sichuan and Guizhou doubled between 1991 and 1992.[175] This observation confirms Yang Zhong's observation of an 'explosion' of government employees at the county and township levels since the late 1980s.[176]

As the following sections will show, the recruitment of personnel was in fact devolved to the local governments after Reform and Opening, but in the early 1990s, the centre sought to repeal some of this autonomy. However, although the formal hiring procedures of government cadres now conformed more to the delegation type, more often than not they were violated. In addition, ad-hoc units could still be formed and temporary staff recruited for the purpose of implementing 'core tasks' specified by the central government.[177]

With regard to these rights and responsibilities, one must distinguish between the recruitment of leading cadres and all other government employees. Promotion and demotion of leading cadres (such as the Party secretary and the head of government at each level) has remained one of the few control mechanisms for the centre after regime ideology had been watered down and the central economic planning system vanished.[178] One crucial mechanism for recruiting leading Party and government cadres has been the nomenclature-system, which put cadre management firmly into the hands of the Party Committees. Until 1984, each level's Party Committee administered the leading cadres of the two subordinate levels. In order to increase its efficiency, however, the administration of leading cadres was decentralized in 1984 by restricting the authority of each level's Party Committee to the subsequent administrative level. While the central government was responsible for the management of 13,000 province- and prefecture-level cadres before 1984, the reform reduced this number to 4,200.[179]

In order to offset this relinquishment of control, the centre implemented stricter regulations for personnel appointment and supervision in 1993 and

1995. In 1993, the Provisional Regulation for Government Employees (*Guojia gongwuyuan zhanxing tiaoli*)[180] was passed to limit and make more transparent public employment (see also the previous chapter). First of all, the regulation stipulated that the total number of cadres employed in the Party and government apparatus was not to exceed the quota authorized by the higher-level government.[181] Based on various indicators such as population size, socioeconomic development and land area, counties were allocated a staff quota that ranged between 350 and 750.[182] In addition, the regulation introduced an element of competition into the selection of public servants, who were to be recruited through public examinations.[183]

In 1995, the Provisional Regulation for the Selection and Appointment of Party and State Leading Cadres[184] decreed that leading cadres should be selected in a four-step procedure consisting of 'democratic nomination', 'evaluation', 'decision on candidacy', and 'nomination and appointment'. To dilute the power vested in the local Party secretary by the nomenclature system, a greater range of actors was to be included especially into the evaluation of nominees, which were proposed by the leaders of the various Party and State organizations, representatives to the People's Congress or by self-recommendation. The evaluation procedure was to be organized by the Organization Department of the Communist Party and included individual talks with leading cadres and other persons related to the candidate as well as 'public assessment' and 'public opinion polls'. As Chen Lijun notes, however, 'public' denoted only a small elite drawn from Party and government organizations, and '[o]rdinary citizens and most government staff are excluded from the process'.[185] Hence, '. . . most local cadres regard the evaluation process as a mere formality'.[186] Since the authority to approve the list of candidates and to eventually fill the vacant position is still vested in the Party Committees, Chen finds 'the nomenclature authority of the Party Committees at al levels . . . still largely intact'.[187]

It follows that the local Party Secretaries continued to wield immense powers over the selection of leading cadres at the subordinate administrative level even after the two Regulations were passed in 1993 and 1995, and professionalism was not always the criterion for selection. Patron-client relationships were equally if not more important.[188] The sanctions for rule-violations were not clearly specified, and violations were punished with an administrative warning, if at all.[189] Leading cadres continued to be appointed by the Party Committee of the higher level, a process in which the Party secretary, other leading cadres and the Organization Department yielded enormous influence. However, both formally and informally, leading cadres at the lower level could exert some influence in the selection process. Formally, they could recommend candidates for Party and government positions, and their opinion was sought in the evaluation process. Informally, they were tied to higher-level cadres by patronage networks. As two high-ranking county cadres have pointed out in interviews with the author, this institutional set-up is responsible for the rampant sale of public positions (*mai guanzi*). One of them, a present county vice-chief, stated that the sale of offices was ubiquitous.[190] The other, a former county chief, claims to have been demoted because he refused to continue paying bribes to the various officials belonging to his and the higher level.[191]

Apart from the 'genuine' cadres which form the 'administrative workforce' (*xingzheng bianzhi*), another, more opaque category of employees existed: the personnel responsible for public services (*shiye bianzhi*). Although the service organizations were largely funded by budgetary revenue,[192] the localities were completely autonomous in staffing them,[193] and the service personnel workforce by far exceeded that on the administrative payroll, as the previous chapter has shown. With regard to these employees, arbitrariness and 'rampant patronage'[194] were hallmarks of the selection process. According to a leading county cadre, 'much more than half' of all jobs on the township level payroll were the result of county-level patronage.[195]

Taken together, a mixed picture emerges with regard to central–local relations in economic management, social service provision and personnel recruitment at the local level. In terms of the rights and responsibilities transferred, administrative decentralization does generally not extend as far as much of the literature on the topic suggests. Clearly, the presence and importance of centrally mandated targets indicates that administrative decentralization has not taken the form of devolution, but rather of the more circumscribed delegation. This is true as well for personnel recruitment although, as we have seen, the centre has so far been unable to enforce these rules.

As for the instruments the centre employs to steer implementation, we illustrated how the curious mixture of elements from both the 'hierarchy' and the 'competition' type was applied to two policy fields. Clearly, a strict hierarchy is the general organizing principle of the Chinese administration, as cadres are grouped into fifteen ranks with different remunerations, rights, responsibilities, and privileges.[196] However, contrary to what one would expect from a hierarchically ordered government, policies are not implemented by means of laws or regulations containing task-specific rules and procedures. Rather, circulars containing the outlines of the given policy and goals to attain are passed down the administrative hierarchy. In this way, the higher level defines a narrow corridor in which competition is to take place. While Linda Li finds this arrangement quite functional, regarding it as a case of local governments '. . . supplement[ing], and even le[ading], central policy in its fine print',[197] Xiaobo Lü is more sceptical about its efficiency:

> Activities to reach certain standards or to upgrade existing facilities often require local funding or, in the case of failure to meet certain standards, monetary penalties for residents, both resulting in peasant payments in rural areas. According to one official report, at one point there were at least forty different kinds of *dabiao* activities in the countryside. This is a major source of corruption by local officials and growing peasant tax burdens in recent years.[198]

Fiscal delegation

The rules for allocating government finances were also a very important factor contributing to an incentive structure that favoured the exploitation of peasants by local governments. As is the case in all other countries in this world, government revenue is composed not only of taxes, but also of various fees, levies, fines, and the revenue derived from government-owned enterprises. The Chinese fiscal system is unique, however, in distinguishing between budgetary (*yusuan nei*), extra-budgetary (*yusuan wai*) and extra-systemic revenue (*zhidu wai*). Budgetary revenue, which equals only around 18 per cent of China's GDP, mainly consists of taxes. Extra-budgetary revenue, as the name implies, is not listed in the central and local budgets. Being mainly derived from State Owned Enterprises (SOE),[199] administrative fees and funds,[200] and local government surcharges on various taxes,[201] it has nevertheless been a crucial source of government receipts, especially at the regional and local levels. Extra-budgetary funds (EBF) are earmarked for specific uses, both in collection and expenditure. They must be listed in designated accounts, and some very rough statistics for the national and provincial levels can be found in the *Statistical Yearbook of China* (*Zhongguo tongji nianjian*) and the *Financial Yearbook of China* (*Zhongguo caizheng nianjian*). The third revenue category is extra-systemic revenue (ESR), which is even more difficult to grasp than EBF. The largest chunk of ESR is the so-called 'self-raised funds', which in industrialized townships mainly consisted of management and contract fees from TVEs. After the budget reform of 1993 (see below), profit remittances levied from enterprises also fell into this category. According to the Regulations on Township and Village Collectively Owned Enterprises, the profit remittances were limited to 40 per cent of an enterprise's net profit, but this limit was often exceeded. In fact, 'the township government effectively controlled 100 per cent of the profits of collective, township-run enterprises'.[202] Half of the remittances went to the township industrial cooperation, which was the township-level office responsible for rural enterprises. The other half entered government coffers, though not as budgetary or extra-budgetary revenue. In reality, however, such ESR went beyond this legitimate portion of profits as '. . . enterprises paid directly for so many of the local government's expenses'.[203] A considerable part of ESR must be considered illegal, and often the term *luan shoufei* (wanton levying of fees) is used coterminous with ESR. Given the fact that this derogatory term denotes a certain degree of non-legitimateness or even illegality, this is not entirely appropriate. As Christine Wong points out, some of these fees are considered legitimate, as they are often used to finance expenditures stipulated by national policy.[204] In addition, receipt and expenditure of such funds had to be reported to the county government along with budgetary and extra-budgetary receipts.[205] In theory at least, townships were not free to use these funds as they pleased, but in practice, they 'were deposited into a general account that the township used at its discretion'.[206]

As Wong, Heady and Woo note, off-budget funds are a 'standard type feature of Soviet-type fiscal systems, created to provide a supplementary source of funds outside of government budgetary appropriations'.[207] 'They are tolerated in recognition of the fact that revenues within the budgetary system are insufficient to finance expenditure responsibilities at the lower levels of government'.[208] Since administrations at the county level and above pocketed budgetary revenue for undertakings which the townships were forced to finance,[209] the latter had to primarily rely on EBF and ESF to finance local public goods.[210] Thus, the flexibility granted to local governments to refine and implement central policies and to hire personnel for this purpose is complemented by a fiscal system that allows the localities to raise revenue in a flexible way and without having to wait for budget allocations.

Again, there are drawbacks to this system that becomes immediately clear when the addressees of such fees and levies are considered. As a general rule, off-budget funds were extracted primarily from those economic actors who contributed most to a locality's GDP. Where TVEs were predominant, off-budget revenue primarily consisted of their profit remittances. The same was true where private enterprises contributed most to the local GDP, but as local governments had far less control over them, they could not be 'milked' to the same extent as TVEs.[211] If neither existed and agriculture dominated the economy, off-budget revenue was mainly derived from fees levied on rural households,[212] making itself felt as the 'burden' discussed in the previous chapter. This was especially problematic at the township and village levels, which had to generate most of their revenue by themselves.[213] Another drawback of this system is that off-budget revenue was not subject to inter-provincial transfers and was thus wholly retained by the spatial unit that produced it. In other words, a large amount of overall government revenue was moved outside the scope of central or provincial redistribution. Before we exemplify this last point by means of a historical analysis, explain the centre's reaction to this problematic development and highlight its impact on the peasant burden, it is necessary to give a brief overview over the taxes, fees and levies extracted from the peasants.

Taxes, fees and levies extracted from the peasants

Those townships that lacked collective enterprises mainly depended on various fees levied on rural households, but also fees levied on contracted farm land, fish ponds, and, if available, from TVEs.[214] Of crucial importance for regions lacking rural industry were two quasi-taxes, the 'village retention' (*cun tiliu*) and the 'township comprehensive fee' (*xiang tongchou*), mentioned in the previous chapter. The village retention financed three expenditures: collective investments at the village level such as streets, bridges and schools; social welfare; and cadre salaries. The township comprehensive levy financed five expenditures: militia training; education; welfare and old-age support; cadre subsidies; and army family subsidies. These levies, collectively known as the 'three retentions and the five unifieds' (*santi wutong*), should, by law, not exceed 5 per cent of the peasants' income from the previous year. The salaries of village school teachers were not included in these levies, so villagers had to make additional payments into a village education fund.[215] Apart from these obligations, villagers had to donate a certain amount of labour for construction and flood prevention each year. However, it had become common practice to have villagers meet their labour obligations by financial means, which created yet another extra-systemic fee. Finally, a multitude of other fees was subject to both geographical and variation in kind, and the amounts levied also fundamentally differed from place to place. Examples are fees levied from TVEs for the purpose of supporting agriculture, financing rural education, and funding social welfare expenditures.[216]

Besides such legitimate exactions, other sources of revenue came from the financial sector. Frequently, townships issued unauthorized bonds or sold off government property, the receipts of which were not recorded. Another very common practice of raising capital was borrowing money from super-ordinate departments or the local branches of the four state banks and other financial organizations, as was shown in the previous chapter. Again, TVEs were a valuable asset for those practices, as township-level governments were not allowed to incur budget deficits. Since bank credits required either sufficient collateral or an 'economic entity' as a guarantor, the township industrial corporation could guarantee for loans by TVEs, whose resources were, as mentioned above, under tight control of the townships. The legal status of the Industrial Corporation benefited this scheme: it was part of the government structure and simultaneously functioned as an economic entity.[217]

Clearly illegitimate and illegal was the creation of funds neither officially sanctioned nor reported to the subordinate level. Since the 1980s, enterprises and villagers alike were plagued by the 'three wantons' (*san luan*), e.g. wanton fees (*luan shou fei*), the wanton creation of funds (*luan jizi*) and wanton apportionments (*luan tanpai*). If the township or the village was in dire need of funds, they would frequently charge made-up and often absurd fees like the 'smokestack fee' or the 'bicycle fee'.[218] As early as 1988, a study conducted in the suburbs of Shanghai identified 188 different fees levied on rural enterprises, but only about one third had been officially approved.[219] Secondly, irregular and non-sanctioned funds were wantonly created, for example for the execution of construction projects which enterprises and villagers had to contribute to. Sometimes, the TVEs were required to directly finance local projects – without their funds taking the detour via the township financial office. Finally, the most common example of 'wanton apportionment' was the practice to apportion among the villagers the cost of magazine and newspaper subscriptions of the village Party branch. Recently, a fourth 'wanton' has been added to the three already existing. Additional revenue was extracted by wantonly fining villagers for absurd or fictitious offences.[220]

The relevance of off-budget revenue

As the subsequent sections will briefly show, off-budget revenue grew to such an extent after Reform and Opening that the central government saw itself forced to intervene. Such assessments must be treated with caution, however. Given that most of such extra-systemic revenue is informal, often illegal and that the lower levels seek to hide them from the central government, it is difficult to determine their magnitude. A good example is a fairly recent statement by then Premier Zhu Rongji at the Fourth Plenary Session of the Ninth National People's Congress (NPC). Elaborating on the plight of China's peasants, he estimated overall farmer's burden at 120 billion yuan RMB 'and even more', making clear that he did not know or did not dare to state the exact figure. One-fourth (30 billion yuan) of the total he attributed to irregular fees.[221]

Fairly reliable assertions can be made about the growth of EBF at the aggregate province level. Figure 3.2 illustrates the mounting significance of EBF by relating them to total government revenue (excluding of course illegal levies, which do not appear in the books).

The figure displays the share of EBF in total receipts at the local level (light grey columns) and central level (dark grey columns) as well as the ratios of budgetary and extra-budgetary receipts to GDP. It highlights three

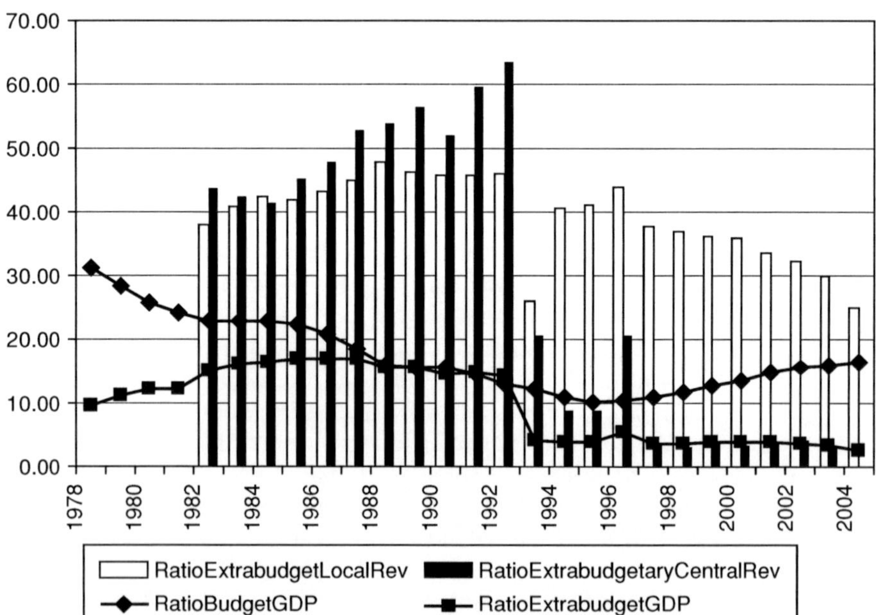

Figure 3.2 Ratio of budgetary and extra-budgetary revenue to GDP, share of extra-budgetary in total local and central revenue, 1978–2005

important issues. First, it visualizes how extra-budgetary revenue increased at the cost of budgetary revenue after 1978, thereby moving an increasing portion of government outside the scope of central or provincial redistribution. It should be added that these tendencies were mirrored at the township level (where a lack of reliable statistics makes such assertions more difficult, however). As figures by the Ministry of Finance reveal, ESR increased from 11.7 per cent to 19.7 per cent of overall township-level receipts between 1986 and 1993.[222] This tallies with a study quoted by Fan Gang, which puts the national average at 19 per cent.[223] With nearly 50,000 township-level governments in China, however, the variation disguised by these averages is probably quite large. For example, EBF in one Shanghai City township visited by Susan Whiting more than doubled and ESF more than tripled between 1990 and 1995, while budgetary funds stagnated.[224] In 1995, budgetary funds only represented 41 per cent of total revenue, while EBF and ESF contributed 15 per cent and 44 per cent, respectively. Similar developments can be observed in Anhui Province, with the crucial difference that off-budget revenue was levied not mainly from TVEs, as in Shanghai,[225] but from the peasants. In Anhui, three major ESR items (village levy, township comprehensive fee, education fund) added up to 56.47 per cent of *county-level* aggregate revenue one year before the Rural Tax and Fee Reform was implemented.[226] This serves to show the degree to which township–level governments came to depend on the peasant burden for revenue.

Second, the overall contribution of EBF to overall government receipts was drastically diminished in 1993. That year, SOE funds were discontinued to be listed as EBF,[227] causing a sharp nominal drop in extra-budgetary revenues. This, of course, does not mean that the government immediately lost all control over the SOE funds and revenues formerly listed as EBF. Rather, they entered government coffers as extra-systemic revenue.[228]

Third, while central government dependence on EBF further declined after 1993, the regions' and localities' dependence rate nearly reached pre-reform levels the following year. This is not mainly, as might be expected, due to the increase in exactions, but a major tax reform in 1994. This reform pushed the centre's share in total budgetary receipts from 22.9 per cent to 55.7 per cent. The corresponding revenue loss at the local level automatically manifested itself in increased dependence on EBF and, one must add, on ESF. On average, local-level EBF grew in the range of 30–40 per cent between 1993 and 1996.[229] Thus, while the centre largely managed to streamline its revenue and expenses into one formal budget, a very large part of local finances was still beyond the reach of higher-level redistribution and formal supervision. In other words, while the 1994 tax reform clearly represents an instance of fiscal recentralization, '[t]he development of off-budget finance in China can be seen as . . . a "back door" decentralization of fiscal control'.[230] This requires some explanation, as it is central for our argument that the roots of excessive exactions from the peasants were systemic in nature.

The 1994 tax reform

Officially, the 1994 tax reform heralded a change of the previous revenue sharing contract systems into a unified 'tax division system' (*fenshuizhi*; often translated as 'tax sharing system'). In an ideal-type tax-sharing system, each administrative level receives a share of all levied taxes. In a tax-separation system, however, each level is entitled to levy certain taxes over whose receipts it fully disposes. Although a variety of tax systems coexisted in China between 1980 and 1988, mixed systems that combined elements of a tax-sharing system and a tax-separation system prevailed.[231] However, neither of these systems had proven satisfactory for the central government, since they did not stop the decline of the 'two ratios' (*liang ge bizhong*), namely the ratio of total government revenue to GDP and that of central to total receipts.[232] Figure 3.3 visualizes the development of the 'two ratios' between 1978 and 2004 on the basis of budgetary revenue (R1) and of both budgetary and extra-budgetary revenue combined (R2). The shaded areas pertains

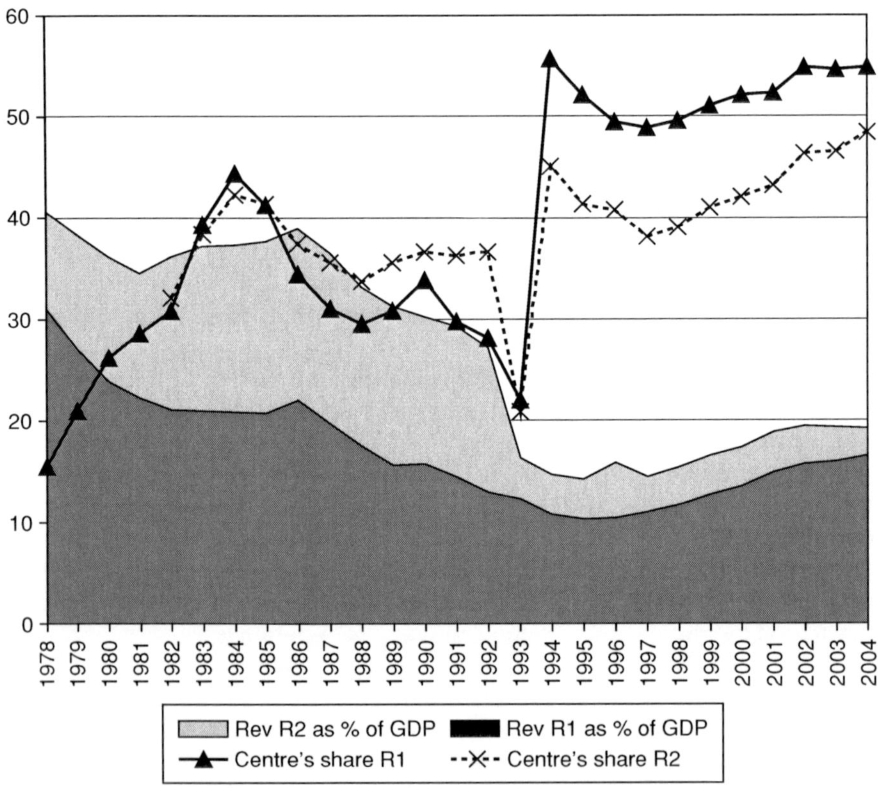

Figure 3.3 Ratios of government receipts to China's GDP, central to total revenue, budgetary revenue and budgetary plus extra-budgetary revenue, 1978–2004

to the overall revenue ratios, while the centre's share in this revenue is represented by the two lines.

As can be seen, the regime's extractive capacity did indeed decline rapidly since the mid-1980s. Obviously, China's tax system, which had been inherited from the planned economy era, had become grossly inadequate to cope with the increased liberalization of China's economy.[233] For this reason, the 1994 tax reform was implemented to strengthen the regime's aggregate extractive capacity, especially that of the central government. Taxes were clearly separated into central, local and shared taxes, but unlike before, the revenue retained from the shared taxes was not subject to negotiations between the centre and each province. Each province had to hand over the same percentage of shared tax revenues, and the centre would then engage in limited redistribution.[234] It was hoped that this version of a hard budget constraint would incite the localities to increase their budgetary revenue.

The new tax system caused the centre's share in budgetary revenue to increase a dramatic 33 per cent. This, however, does not mean that the centre had more to spend than before. Part of the deal between the centre and the provinces that brought the new tax system into existence was the promise that provincial revenue would not be allowed to decrease as a result of the reform.[235] Accordingly, it was clear from the beginning that fiscal transfers would be an inherent part of the new tax-division system. If the fiscal transfers between the centre and the localities after tax collection are netted out, the centre's share in budgetary revenue did not increase after the reform, but kept decreasing. Figure 3.4 illustrates this. In 2005, the centre's share in government receipts had reached an all-time low, no matter which calculation is applied. The share in budgetary revenue had declined to 18.29 per cent, and if EBF are considered as well, the share is even lower (16.34 per cent).

This indicates that it was not the centre's fiscal capacity that had been strengthened as a result of the reform, but rather its redistributive capacity. However, this does not mean that the fiscal situation at the local levels was improved. The following paragraphs will show that the contrary was true.

A close examination of the document underlying the reform[236] reveals that the tax reform at first applied only to the fiscal relationship between the centre and the provinces. As for fiscal relations at the sub-provincial level, the circular merely states that 'all people's governments of the provinces, the autonomous regions and the cities directly under central administration must set up the budgetary system prescribed by this regulation'. More concretely,

> [d]uring the implementation of the tax division system, all regions are required to carry out research, in good time sum up the outcomes and solve emerging problems in order to gradually fulfil the preconditions for improving and perfecting the budgetary administration of the tax-division system.[237]

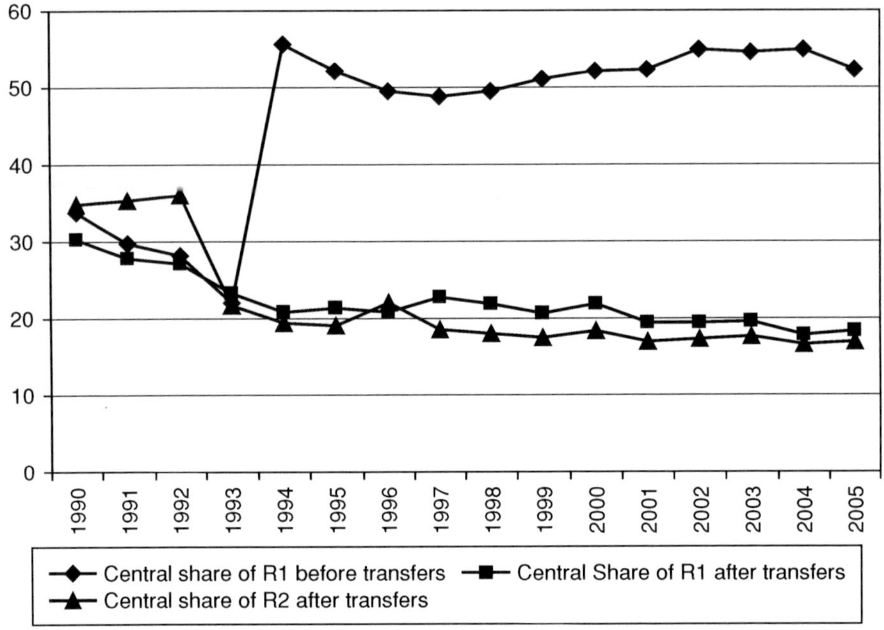

Figure 3.4 Centre's share in budgetary revenue before and after transfer payments, and budgetary plus extra-budgetary revenue after transfer payments, 1990–2005

As can be seen from these short quotes, competition under hierarchy also applied to the implementation of the 1994 reform at the sub-provincial levels. The centre prescribed the outcomes to be achieved regardless of the contexts the individual provinces and cities might have found themselves in, and left it up to them to devise and implement polices conducive to reaching these aims. This does not mean that central-level policy-makers thought that no problems would occur, but again the localities were required to solve these problems by themselves and provide good examples for others. As a World Bank study points out, 'the freedom of local governments for setting the rules for lower levels is constrained only by the ability of the lower levels to appeal for intervention from the central government in egregious cases'.[238] For this reason, a variety of revenue assignment systems continued to exist at the sub-provincial levels despite repeated exhortations of the centre to implement the new system at all levels and everywhere. Only twelve provincial-level governments followed these rules,[239] while the others kept part or all of the revenue that would have accrued to the sub-provincial levels according to the new system.

Therefore, a peculiar system emerged: while the centre increased its redistributive capacity vis-à-vis the provinces, the same is true for the provinces, the cities and the counties vis-à-vis their respective subordinate levels. In practical terms, and with regard to budgetary revenue, this means that the 1994 tax reform made each level more dependent on fiscal transfers from above than before. Seen from another perspective, this also means that each level's fiscal control vis-à-vis the subordinate level had increased. In fact, with the concentration of local tax revenues now in the hands of the province, tax sharing was turned down side up. Whereas resources were levied at the grassroots level and contracted upwards before, they were now allocated downwards. Before, the budgetary revenue shortage affected primarily the central government, now it affected the local and grassroots levels. However, expenditure responsibilities were not adjusted accordingly. As shown above, the setting of collection and construction targets even intensified after the reform.[240] Naturally, governments at the county and township levels tried to compensate this shortage by increased collection of EBF and ESR, and by borrowing money where still possible.[241]

Townships and towns were the tax collecting agencies of the counties, so they bore the brunt of the 1994 reform. Those that did not have collective enterprises mainly had to rely on fees and levies which they extracted from the rural populace.[242] This strongly suggests that the 1994 tax reform brought about an inverse relationship between administrative level and autonomy over budgetary revenue. The lower the level, the less budgetary autonomy it enjoyed. At the same time, however, villages, townships and counties were hard pressed to come up with resources to finance increasingly high achievement targets. The only solution was to compensate the decrease in budgetary revenue autonomy by an increase in 'informal revenue autonomy'.[243] In other words, the part of local revenue that consisted of off-budget levies, funds, fees and bank credits increased steeply at the expense of formally sanctioned fiscal revenue. Both 1994 and 1995 saw a 30 per cent increase of EBF over the previous year. In 1996, they even increased 62 per cent.[244]

In July 1996, the State Council passed a document to rein in EBF. For the first time, their scope was clearly defined, and the document stipulated that they had to be deposited in designated bank accounts. Finally, some fees were turned into taxes.[245] The move was justified on the grounds that 'funds had been intentionally withdrawn from central government purview and redistribution by being listed as EBF and not as budgetary revenue'.[246] The move seems to have been successful, since the relevance of EBF has decreased steadily after 1996, as Figure 3.2 shows.[247] At the national level, they equalled merely 17.8 per cent of budgetary revenue in 2004, an all-time low since 1972.[248] Given that the squeeze in budgetary revenue had forced local governments to pursue revenues in the off-budget sector, it can be expected that the retrenchment of EBF was compensated by an increase in ESR. Very likely, this was one of the main systemic factors leading up to the intensification of the 'burden problem' in non-industrialized regions.

Political decentralization

In no dimension of the centralization–decentralization continuum is the difference between formal and informal rights and responsibilities in China more pronounced than in political decision-making. As explained in the introduction to this chapter, the sub-concept of political decentralization places a premium on decision-making powers of regional or local governments empowered by direct and competitive elections. In China, elections for People's Congresses at the national, provincial and city level are both indirect and non-competitive, while elections for county- and township-level assemblies are direct, but equally non-competitive.[249] Since the representatives at all these levels are in fact appointed, the concept cannot be applied here. Elections for Villagers' Committees (*cunmin weiyuanhui*; VC), however, can be considered at least semi-competitive, even if the character of the election and the level of competitiveness are subject to large spatial and temporal variation.[250] Still, it is rather difficult to make a clear case for 'villagers' autonomy' (*cunmin zizhi*) as an instance of political decentralization, none the least because the Communist Party branch as the 'leadership core' of village government can overrule VC decisions,[251] and because the villages are not a formal level of government.[252]

The picture changes, however, if the analysis moves beyond examining the formal institutions that structure central–local relations and examines whether they are substituted, complemented or contradicted by informal rules of the game.[253] Kenneth Lieberthal, for example, was among the first to question the centralization and coherence of Chinese politics. The well known 'fragmented authoritarianism' concept stresses the importance of 'negotiations, bargaining, exchange, and consensus building'[254] in a polity that is territorially and functionally intersected, and where 'no single body has authority over the others'.[255] In fact, central control was never total, not even during Mao's reign,[256] and China's transition from a totalitarian to an authoritarian system and from a planned to a marketized economy further loosened central control and widened local spaces for political manoeuvring. Leaders made far less use of coercion and ideological campaigns to bring subordinates into line, the creation of off-budget revenue has reduced dependence on the higher levels, and local functional units have become accountable not only to the corresponding higher-level functional unit, but also to their own government.[257] Hence, political decentralization has led to the emergence of a 'double identity' of local governments, which increasingly represent not only the interests of the centre, but also that of their respective constituencies.[258] As Thomas Heberer notes, 'the central government is not anymore the sole instance of decision-making. Rather, provinces, local governments, townships and villages are political actors who pursue their own interests'.[259]

However, there are some clear drawbacks of political decentralization in China, especially where it is perverted to local state capture. Jae Chung Ho has convincingly shown that the constellation of actors in policy experiments is crucial for their success. More often than not, central and local actors are connected by a chain of clientelist ties.[260] While clientelism can work for the common good, it often takes the less desirable forms of factionalist strive[261] and rampant corruption.[262] In addition, political decentralization in economic decision-making has also led to an increase in regionalism, communalism, and localism. According to Thomas Heberer, local protectionism, internal strife for resources, the independent conduct of foreign economic relations, and deviation from central policies are indicators of these developments.[263] Furthermore, the significant degree of political decentralization has also been manifested in the increased number of policy circulars passed at all administrative levels. According to Xiaobo Lü,

> this has led to a situation where more and more decrees, directives, and ordinances are issued and yet are becoming less and less effective, while the jurisdiction of the decrees becomes narrower and more limited, facilitating regionalism and conflicting rules.[264]

As a result, existing parameters of social relations were replaced in some localities after Reform and Opening started, and the central government found itself unable to regain dominance. Many of the contemporary problems such as the persecution of minority group leaders, village-level dictatorships, mining accidents, organ harvesting, the violation of copyrights and of course the burgeoning corruption problem confirm that state capture by local officials is widespread. As emphasized throughout this study, however, we can expect some regional variation in the relationship between the central and local states. In the realm of economic policy-making, for example, Richard Baum and Alexei Shevchenko have discerned four different types of local state interference. While some cadres clearly are predators, there are many others who actively develop the local infrastructure to create a beneficial investment climate. Yet others make use of personal relationships to syphon off rents or benefits, or engage into business themselves.[265]

In sum, local officials can have considerable autonomy in shaping local politics. This autonomy can be the result of tacit acquiescence or even the informal delegation of power by central government politicians, and policy innovations and successful experiments at the local level can influence the formulation of national policies and contribute to the fine-tuning of existing ones. In such cases, central–local political relations are not zero-sum games, but situations of mutual empowerment. On the other hand, there are many instances where the central level has proven unable to steer developments at the grass roots and local level. In such cases, decentralized policy-making is not the result of tacit acquiescence, but of a loss of control. Hence, a multi-faceted picture of central–local political relations emerges. As has been shown, they display great variation across space, time, and issue area, and formal political centralization coexists with both top-down and bottom-up informal decentralization.

Conclusion

To sum up, after Reform and Opening local governments enjoyed what could be aptly called 'autonomy until cancelled'. The centre delegated many administrative tasks that had formerly been provided by the communes, and gave the localities considerable freedom to hire personnel and to generate revenue. They were given targets to achieve, but enjoyed much autonomy in policy implementation, revenue raising, and even in economic and social policy formulation. Still, they were accountable to the centre, and the centre could selectively curtail their autonomy. As we have seen, administrative delegation facilitated a mode of political steering which we termed competition under hierarchy. This institutional set-up proved beneficial where the conditions were right, and had negative consequences where they were not. What exactly these conditions are remains to be seen, but the findings so far suggest that competition under hierarchy works well for local government that are positively imbued with natural, geographic or financial resources. In contrast, remote rural areas depending on agriculture seem to fare less well. In order to conform to the economic and social standards set by the centre, they were forced to raise funds outside the budget, which was easy in a fiscal system that puts a premium on off-budget revenue. Lacking rural industry and having lost their creditworthiness, however, they became forced to exploit the peasants for revenue.

Nevertheless, a part of this revenue was certainly used to line the pockets of those working in and for the local administrations. With regard to central–local relations in the realm of political decision-making, party discipline and ideology were found to have been replaced by informal networks that pursued their own agendas. In what represents a crucial instance of bottom-up political decentralization, the centre increasingly lost control to influence politics at the county level and below, resulting in the creation of 'dukedom economies' (*zhuhou jingji*) and 'independent kingdoms'. One visible result of this constellation was the rapid decline of the centre's share in total government receipts. The centre reacted by implementing a 'tax separation reform' and recentralized much of the formal revenue, which, however, forced especially the poorer localities to increase illegal levying.

To conclude, is not difficult to see how this multi-faceted mix of competition under hierarchy, political and fiscal (re-)centralization as well as local state capture produced both the institutional channels and the incentives for extracting moneys outside the budgetary channels. It has been shown that rural enterprises were the chief objects of such exactions, and that illegal bank credits provided another revenue source. However, where rural industry was absent and the credit limit exhausted, China's peasants provided another crucial revenue source.

4 The limited agendas: the peasant burden and rural fiscal reform

The previous chapter sketched the complex institutional framework of intergovernmental relations in China, and showed that these structures both incited and enabled local cadres to extract funds from the peasants. Against this backdrop, this chapter examines how far the public discourse and official interpretations of the peasant burden problem acknowledged these deficiencies. This is important because a policy can only work if it is based on realistic assumptions. The following sections examine when the peasant burden entered the political and public discourse, how intensely it was discussed, how the issue developed over time, and which policies resulted. The peasant burden discourse is contrasted with the fiscal reform discourse to find out how the former influenced the latter and when they began to overlap. Of course, agendas do not develop of their own volition, but are the result of argumentative and power struggles waged by various groups of actors. This chapter examines the roles of central and local politicians and bureaucrats as well as peasants, scholars and journalists in the agenda-setting process. As will be seen, the RTFR commenced in a bottom-up manner, which was characterized by decentralized innovations conducted in innovation networks spanning several administrative levels. Although the pioneers of rural reforms certainly hoped that their actions would reap political rewards, the framework within which they cooperated was rather one of co-operation than of competition under hierarchy.

Narrating the peasant burden

In order to get a comprehensive understanding of the peasant burden discourse, the present section will provide a quantitative as well as a qualitative analysis of the issue's development. A very simple quantitative analysis is conducted on the base of three comprehensive online databases, namely of the government mouthpiece, the *People's Daily* (*Renmin Ribao*; henceforth: PD), the China Academic Journals database (*Zhongguo qikan quanwen shuju ku*; henceforth: CAJ), and the 'National Index of Chinese Newspapers and Periodicals' (*quanguo baokan suoyin*; henceforth: 'National Index') hosted by the Shanghai Library.[266] While the PD archive and CAJ are full text databases, the National Index only carries the headlines of newspapers and some magazines. The PD archive, which at the time of writing covered the years 1946 to 2006, was chosen to capture the government's role in opinion formation, while the National Index reflects tendencies in public discourse, and the CAJ in scholarship. The National Index spans the period from 1857 to 2006 and contains 15 million references, mostly from national and regional dailies.[267] CAJ provides 22 million full text articles of 8,200 academic journals mainly from 1994 to the present, but it seems to have started to cover earlier years as well.[268] Whereas the PD archive is a complete collection of all its articles, the National Index and CAJ are probably not comprehensive. However, their large size nevertheless makes them representative samples.

The aim of this exercise is to establish the intensity with which, over time, undue exactions from China's peasants were discussed in the Party mouthpiece and in newspapers and journals, and to find out how what problems were identified and what solutions were suggested. I use the frequency of certain key words in newspaper and journal headlines as the main indicator for the social, scholarly and political relevance of the issue. The quantitative results will then be contrasted with a qualitative analysis of the agendas' development. Since 'peasant burden' (*nongmin fudan*) is the term nearly always used when referring to excessive exactions from China's farmers, its frequency is a very reliable indicator for the changing intensity of the agenda. As for the development of the rural fiscal reforms agenda, I searched for headlines that contained the two keywords 'tax' (*shui*) and 'fee' (*fei*). These keywords proved to be reliable indicators because they capture all terms used in the reform discourse (Tax-for-fee (*feigai shui*); Rural Tax and Fee Reform (*nongcun shuifei gaige*); reform of the rural tax and fee system (*nongcun shuifei tizhi gaige*)). Nevertheless, I minimized the chance of missing out important contributions subsumed under different headings by testing for related keywords such as 'rural reform' (*xiangcun gaige*, *nongcun gaige*) or 'fiscal system reform' (*caizheng tizhi gaige*). Figure 4.1 visualizes the frequency with which the keywords appeared since the founding of the PRC.

Remarkably, the frequencies with which the term was mentioned in the PD and the National Index correlate significantly until 1995.[269] Most probably, the national media had to follow the trends set by the Party's mouthpiece. The fact that both hardly mentioned this contentious issue during the 1987–89 political crisis seems to confirm this assumption, as does our knowledge about media control in the PRC.[270]

The template

There were a few references to the peasant burden during the Mao era, but the problem was really brought to public attention only at the beginning of the Reform and Opening Period. For 1978, the NI registers more than 100 articles on the peasant burden. Clearly, this peak was prompted

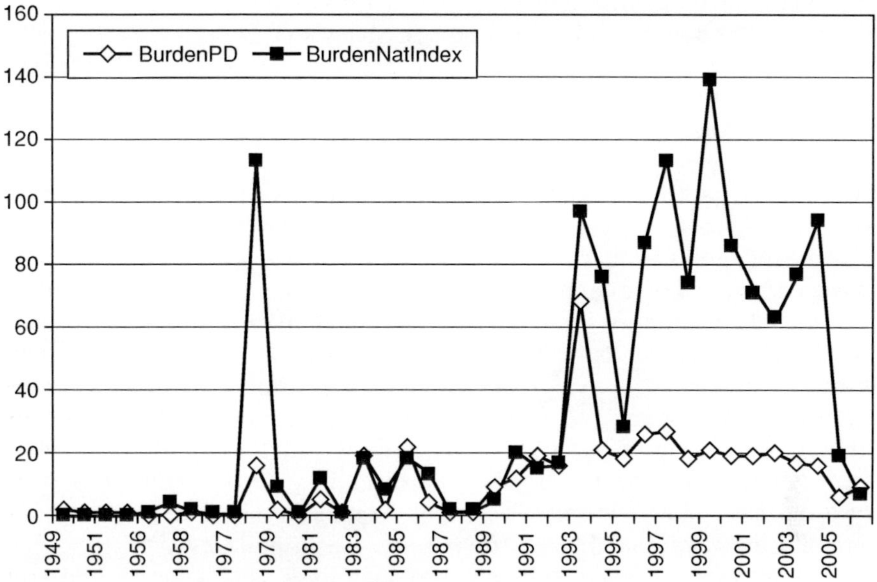

Figure 4.1 Frequency of headlines mentioning 'peasant burden' in *People's Daily* and the National Index, 1949–2006

because Beijing chose to highlight the issue. In early 1978, the Central Committee passed a circular demanding the reduction of exactions from the peasants,[271] which was followed by a PD front-page editorial headlined 'Implement the Party's Policy, Reduce the Peasant Burden' (*luoshi dang de zhengce, jianqing nongmin fudan*) on July 5, 1978. The editorial not only set the tone for future documents, but also for the ensuing newspaper articles on the issue. Hence, the 'public discourse' on the peasant burden problem largely converged with this official interpretation. Immoral and greedy local cadres are made responsible for bad government, but structural deficiencies are not mentioned. As a consequence, the Central Committee promises to defend the hardworking and hapless peasants against its own agents, over whom it has lost control:

> [W]e must start to mobilize the masses against the class enemies supported by the Gang of Four who savagely destroy the socialist ownership system and the collective economy of the people's communes, and against these corrupt thieves and profiteering criminals who greedily swallow up the fruits of people's labour. In addition, we must . . . resolutely intensify the attacks to recover those ill-gotten gains and stolen property. At the same time, we must resolutely stop the evil bourgeois wind blown up by the Gang of Four and strictly forbid extravagant banquets, invitations and present-giving, extravagance and squandering [public resources], and the wanton construction of public

> buildings and offices. From now on, those who use public funds for wining and dining and for giving presents will be required to repay these funds, and all wantonly constructed public buildings and offices will be confiscated by the state. Those who commit serious errors will be held responsible by the Party Disciplinary Committee and State Laws.[272]

As will be seen, the substance of the government's narrative espoused in the editorial has not changed much until today; only the wording has become less drastic, and the macro-force that tempted local cadres to abandon 'the advanced nature of a communist' has changed from the ominous 'Gang of Four' to the temptations offered by an unbridled market.[273] The root causes of the peasant burden have remained the same ('overspending and embezzlement by several collectives', 'all sorts of apportionments and subsidies [taken from the peasants]', 'demanding money from the production team at every opportunity'), as have the exhortations ('put Party policies into practice, reduce the unreasonable peasant burden, mobilize the enthusiasm of the peasant masses towards Socialism'), and the remedies ('reduce personnel and simplify the government structure, reduce . . . expenditures').[274]

In short, the peasant burden is not attributed to systemic deficits, but mainly to the moral degradation of the local leadership. Exhortations to 'sincerely change ways and work-style', and the threat that 'those who seriously violate the law and Party discipline must be severely punished' are repeated in today's documents almost to the letter. Likewise, leaders at the higher levels have always been exhorted to take the leadership and 'personally investigate the peasant burden problems and think about what can be done'.[275]

As Thomas Bernstein points out, however, the main purpose of this early peak in reporting about the peasant burden was not to address the problem itself, but to instrumentalize it for mobilizing public support against the enemies of the reformist faction.[276] As a consequence, the issue quickly vanished from the public agenda after Deng Xiaoping had consolidated his power in late 1978, but reappeared now and then. Articles in the early years of Reform and Opening seem to have been mainly prompted by political deliberations,[277] and they bore no hints that peasant collective action against excessive exactions had taken place at that time. Thomas Bernstein indicates that peasants started to express their dissatisfaction with excessive exactions only in the second half of the 1980's, soon after the regime's focus had changed from stimulating agricultural production to promoting rural industrialization. Bernstein notes that '[t]he Beijing journal *Liaowang* depicted peasants as being in a "rebellious state of mind" (*zhaofan xinli*) in the Winter of 1986–87 in reaction to "extreme measures" taken by local officials to compel fulfilment of state sales quotas.'[278] In 1988, the *Farmers' Daily* (*Nongmin Ribao*) reported to have 'received more than a thousand letters a day, mostly complains against 'village tyrants.'[279] Also, first reports of rural riots appeared[280] but quickly subsided again as anti-regime student demonstrations intensified in the provinces throughout 1987.

The official interpretation of peasant burden

The second peak of the peasant burden headlines after 1978 fell on the year 1993. This peak coincides with the 'Renshou riots', a major mass protest staged by peasants in Renshou County, Sichuan Province, to protest local officials' disregard of the centre's newly passed regulations on the peasant burden.[281] Important circulars demanding the reduction of exactions from the peasants had been passed since early 1990. It was decreed, for example, that the *santi wutong* 'should in general' not exceed 5 per cent of average peasant income.[282] Furthermore, the levying of fees, funds and apportionments not sanctioned by national laws was forbidden.[283] The 'Regulation Regarding the Administration of Fees and Compulsory Labour Shouldered by the Peasants' passed in December 1991 further specified these rules[284] and, uplifting them to the status of a Regulation, gave them more authority. Nonetheless, another Notification reiterated the importance of the Regulation in mid-1992.[285]

Bernstein and Lü's thorough account of the Renshou riots highlights a capturing sequence of events. The riots only started after the peasants had learned about these documents, but they caused the passage of even more documents. Peasant unrest had been flaring up during the first months of 1993, but they turned into 'large-scale riots' in late May and early June.[286] While the conflict was brewing in the first half of 1993, five central-level circulars were passed, one of them labelled 'urgent'. Except for the agricultural tax and the *santi wutong*, all other exactions were temporarily frozen.[287] Ironically, a PD article praised Renshou's cadres for 'listening to the farmers' and 'reducing burden' at the very height of the riots.[288] Very shortly after the riots had subsided, the PD published a series of articles titled 'Perspectives on the Peasant Burden Problem' which were prominently placed on top of the second page on four consecutive days. The series elaborated on 'Which burdens are legitimate, and which are not?',[289] 'Construction costs may not be apportioned',[290] 'Solve problems from the peasants' perspective'[291] and 'Where to start to reduce the burden?'.[292] The peasant burden made the headlines of twenty-one *People's Daily* articles in June, which is much more than had usually been published during a whole year! However, it was never mentioned that riots had occurred. Twelve more articles appeared in July that year. After that, the centre obviously decided to lessen public attention to the burden problem. No articles were published in the subsequent three months, only one in November and three in December. Nevertheless, five more circulars appeared in the second half of 1993, and the 'Law on Agriculture' was revised to give peasants the 'right to refuse' unauthorized fees, fines, apportionments and compulsory fund-raising.[293]

In 1994, the number of peasant burden headlines returned to their pre-riot intensity, but from 1996 on, their number soared again. That year, another batch of circulars was passed. As outlined in the previous chapter, local governments had increased irregular levying after the 1994 tax reform had imposed hard constraints on their budgetary revenue. The 1996 circulars, in turn, sought to keep the growth of irregular levies in check. The passage of these circulars again prompted burden-related protests at a 'substantial scale',[294] and even more regulations and peasant burden headlines followed in 1996 and 1997. At least twenty-four central regulations addressing the burden problem had been passed between 1985 and 1996 alone, but by 2001, the number had exceeded seventy.[295] Peasant-burden headlines peaked again in 1999, the year the implementation of a rural fiscal reform was proclaimed by the central government, but their number decreased until 2002, when it was finally decided to implement the reform nationwide. After the agricultural tax had been rescinded in 2004, the number of headlines dropped again, and by 2006, the issue seems to have nearly vanished from the agenda.

As these sections show, there is an almost cyclical relationship between the passage of documents that seek to limit the peasant burden, burden-related protests, reporting on the peasant burden and the passage of further documents. Two related questions are of interest at this point of the study. How did these documents and news reports explain the burden problem, and what counter-measures did they prescribe? Was the discussion to engage in rural fiscal reforms part of this agenda?

The general content of the burden-related circulars is succinctly summed up by Thomas Bernstein and Xiaobo Lü:

> They acknowledged the failure of previous directives to reduce burdens or to prevent burdens from 'bouncing back' (*fantan*) after they had been cut. They diagnosed the problem as very serious. They exhorted officials seriously to deal with the problem and to put generalist leaders, especially the territorial Party secretaries, in charge of the matter. They acknowledged that officials needed to raise their level of 'ideological understanding' of the burdens problem, a stance that reflected a general tendency to attribute lack of compliance not to conflicts of interest but to lack of understanding of the real situation. According to this view, officials must realize that peasants were still poor and their capacity to 'endure' was limited.[296]

As expressed above, burden-related documents tended to attribute excessive exactions to local leadership failure ever since Reform and Opening commenced in 1978. Taking the stance of 'the peasants' and neglecting to address the systemic roots of the burden problem, the centre obfuscated its own role in generating the burden problem and instead played peasants and cadres out against each other. Media reports generally followed this line. For example, in an article 'observing and pondering the bouncing back of the peasant burden', the *Farmers' Daily* (*Nongmin Ribao*) complains that '. . . some people believe that because they belong to departments or local governments, they hold all power in their hands, and they will dare anything to collect fees'. As the 'deeply rooted causes for the swift and violent bouncing back of farmers' burdens' the article identifies

> the soaring growth . . . of 'self-raised' and extra-budgetary revenues and expenditures; the sharp expansion of local personnel contingents; the malignant inflation of individual and small group private interests; the escalating competition for extravagance; the problem that 'for every policy at the top, there is a counter-policy at the bottom'; as well as such things as the lack of effective safeguards for the legal rights and interests of farmers.[297]

Similarly, after a detailed and graphic description of the 'truly onerous' and 'bizarre' fees peasants have to pay, an article in the *Legal Daily* (*Fazhi Ribao*) on the 'multiple causes' of peasant burden complains:

> Despite the abolition long ago of the demand that peasants contribute money, materials, and labour to help reach certain targets, certain localities still continue that practice, extorting money and goods from peasants on all kinds of pretexts. The CPC Central Committee has banned some activities by formal decree and ordered that others be firmly corrected, including the compulsory apportionment of expenses. Be that as it may, a number of localities has not put a stop to them or corrected them.[298]

Although the lengthy article recognizes that 'many of the things discussed above involve the system and cannot be resolved overnight', it concludes that 'we have reason to criticize some individuals and departments for failing to carry out and realize the Party's and government's rural policy properly, and to regard burden reduction as the political priority that it is and work on it accordingly.'[299]

Cheng Li convincingly asserts that such reporting is part of a 'strategy to localize the social unrests and blame local leaders.'[300] Similarly, both Jae Ho Chung and O'Brien/Li argue that the limited and carefully circumscribed freedom to report on such issues strengthens the monitoring capacity of the centre (while putting the blame for these occurrences on local officials, it should be added).[301] As Li Lianjiang has found out, the centre has been quite successful in generating trust for itself by deliberately distorting information and blaming policy failures on deviant local cadres:

> Scholars have also found that the mass media significantly affect the rise and fall of political trust. In China, oddly enough, it is perhaps the lack of free media that enhances villagers' trust in Beijing. . . . The Centre, in effect, runs a non-stop campaign for public support with its gigantic propaganda machinery, and one tactic for boosting public confidence in it is to occasionally condemn local officials for disobeying the Centre

and mistreating the people. Villagers hear many beautiful promises from central leaders; it is no wonder that many of them come to believe in the Centre and its policies.[302]

The peasant burden and fiscal reform

In these one-sided accounts of the 'burden problem', differentiated analyses of the systemic roots of the peasant burden and suggestions for rural fiscal reforms were absent. In fact, rural fiscal reforms entered the public agenda only in 1997, as Figure 4.2 shows.

Figure 4.2 contrasts the number of headlines containing the words 'peasant burden' with those carrying both 'tax' and 'fee'. Since rural fiscal issues were not headlined before 1988, the figure only visualizes the time span between 1988 and 2006.[303] The lines refer to the coverage of the peasant burden in the three indices, while the number of headlines containing the words 'tax' and 'fee' is represented by the grey surfaces.

As shown above, the 1993 peak in peasant burden headlines indicates that the Renshou riots were considered a major issue by the central government,

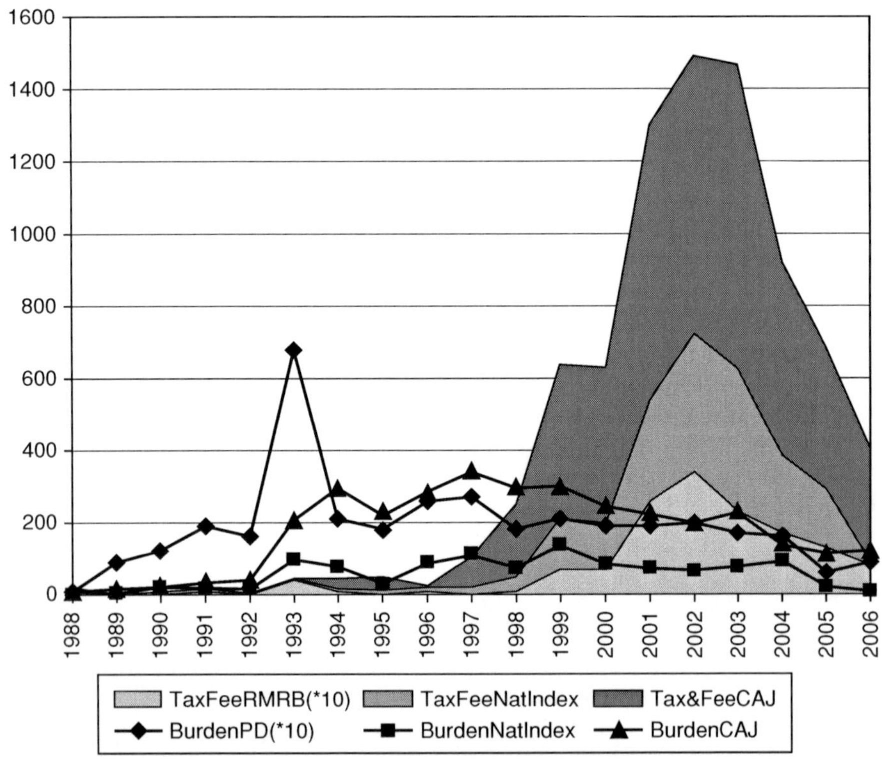

Figure 4.2 Headlines mentioning 'peasant burden', 'tax' and 'fee' in the *People's Daily*, the National Index and China Academic Journals, 1988–2006

especially because they immediately followed the central government's passage of burden-reduction regulations. However, almost all articles published during that peak prescribed measures to reduce burden, reprimanded local governments for their lack of enthusiasm in cutting exactions from the peasants, or portrayed successful examples of burden reduction. One would suspect that such a moment of high tension provided the ground for rural fiscal reforms to enter the agenda, yet as Figure 4.2 shows, this was not the case. There are several possible reasons. First, the relevant departments and Ministries in the centre and the provinces were too occupied with the preparation of the 1994 tax reform. Second, the centre might have hoped that the tax-separation system would somehow also solve the burden problem. After all, one effect of the reform was an increase of resources at its disposal for redistribution. Third, the centre might have hoped for a late success of its newly passed anti-burden regulations, and might have found a policy change at this point of time immature. Fourth, the topic might have been given less priority than the documents and the reports make believe.[304] Fifth, rural fiscal reforms might have entered the agenda, yet not the media. These assumptions are not mutually exclusive, and as has been shown so far, evidence exists to confirm each of them. Of particular interest as to the relationship between the peasant burden and fiscal reform agendas is the fifth assumption, which will be followed in more detail. As the following paragraph and the next section will show, rural fiscal reforms had been on the political agenda since 1988, but entered the public agenda only after experiments had been well under way.

The grey areas of Figure 4.2 indicate that large-scale media and scholarly coverage of rural fiscal issues commenced only in 1997. As the next chapter will show, however, fiscal reform experiments had already been conducted in more than fifty counties in central China that year. Obviously, reporting on these bottom-up experiments was forbidden, which means that they were not publicly discussed. Again, all three sources display similar tendencies. Tax and fee headlines increased when the dispersed experiments were made public (1997), when it was decided that a Rural Tax and Fee Reform would be implemented in one full province (1999), and when the Anhui experiment was evaluated and the reform became a national policy (2001 and 2002). Despite the dramatic developments that followed, interest subsided thereafter.[305] To conclude, rural fiscal reform issues entered the discourse just after the reporting on the peasant burden had gradually subsided. In fact, there seems to be an almost inverse correlation between these issues. If the attention that the peasant burden gained after the Renshou riots and other incidences did not put issues of rural fiscal reforms on the public agenda, where did the RTFR originate?

The rural fiscal reform agenda

Many accounts of the RTFR follow the official narrative that the policy originated in early 2000, only shortly before its trial implementation in

Anhui Province. Furthermore, the official narrative posits that 'the difference between the RTFR and the Household Responsibility System is that the latter originated from below, whereas the RTFR came from above.'[306] As the following paragraphs will show, this is not entirely true. In fact, a network of specialists and politicians whose aim was to reform rural finances had started to form as early as 1988. By early 1991, former Premier Li Peng had explicitly, though not formally, endorsed such experiments. Still, it took another two years before very small-scale experiments began to be carried out, and another nine years before the policy was implemented nationwide.

The following paragraphs are a fascinating account of how the shifting constellation of very diverse factors such as the commitment of individual actors, personal relations, competing organizational interests, high-level power politics and environmental disasters opened and closed windows of opportunity. They show that the agenda-setting process can be divided into two stages. First, rural fiscal reforms became an item on the central leadership's agenda as early as 1988, but were eclipsed by the regime crisis in early 1989. They were reintroduced in early 1991 and stayed on the centre's agenda ever since. Despite then Premier Li Peng's tacit approval, however, it took more than two years until limited experiments were implemented. Alliances spanning all administrative levels had formed, and they were powerful enough to resist the implementation of such experiments even against the explicit approval of a very powerful faction in the central leadership and leading cadres at the provincial, district and county levels. The second stage began after this resistance had largely been overcome. A variety of fiscal reforms was implemented first in two townships and then all across northern Anhui. Shortly thereafter, the central government's official endorsement of these experiments heralded the beginning of the policy formulation phase at the national level. These two stages will now be described and analysed in detail.

Bottom-up development

Basically, the process started in 1988 and with a personal connection. The main protagonist of this stage was He Kaiyin, a specialist on agricultural issues and high-level advisor to the government of Anhui province. He was an ardent champion of the Household Responsibility System, and had been a protégé of Wang Yuzhao, the former governor of Anhui Province. One of Wang's responsibilities, who had by 1988 become the vice-head of the Central Committee's Rural Politics Research Office, was the formulation of new ideas for improving China's rural development. That same year, Wang solicited ideas from He, who avidly responded to the challenge and wrote an article titled 'Where is the Path Towards a Second Reform Step in the Countryside?' (*Nongcun dier bu gaige de chulu he zai?*).[307] The article championed to grant the peasants long-term leases on their land, to reform the agricultural tax and to reduce peasant burden.

Specifically, he proposed to levy the agricultural taxes in kind instead of cash. The tax rate would be 10 per cent of the average yield per *mu*[308] land of the previous three to five years, and the yield average would be re-established every ten years. The *santi wutong* would be collected along with the agricultural taxes, also in kind, and set at 5 per cent of the average harvest per *mu* land. Having submitted this amount of grain, the peasant's obligations to the government would be fulfilled. All other levies would be illegal.[309]

The advantage of this system was that by paying their dues, the farmers would automatically help the township governments fulfil their grain purchase quotas. This would make higher-level subsidies for grain purchases to ensure China's food self-sufficiency unnecessary. A further plus of this policy was that the peasants would not have to sell their grain to grain procuring stations at lower than market prices. In fact, the two-tier grain price system could be totally abolished, and the peasants could sell all their grain at the higher market prices. Also, by paying their taxes and fees in kind, peasants would not be subject to 'hidden burdens'. In the grain stations, it was quite common to extract more grain than officially required, underreport the amount received, and pay farmers even less than the state price. Often, only IOUs were handed out. It was common knowledge that the grain stations were a hotbed of corruption, much to the detriment of China's farmers. Finally, the tax on agricultural special products would also be levied in kind, and the cash crops sold by the government for a profit. This profit, He argued, could be '. . . used to strengthen the agricultural sector by mending irrigation facilities, developing agricultural mechanization, improving agricultural technology, and establishing agricultural development funds'.[310]

He Kaiyin sent his suggestions to the CC's Rural Research Office and the State Council's Rural Development Research Centre, but the window of opportunity had already closed. In 1989, the CC's research institution was dismantled, and the State Council's subordinated to the Ministry of Agriculture. In addition, the post-Tiananmen struggle between Conservatives and Reformers created an atmosphere quite hostile to innovations.[311] Nevertheless, in the beginning of 1990, He submitted a second article to the Anhui office of the government news agency *Xinhua*. The article was included in the agency's 'Internal References' (*neibu cankao*), and a special issue of the *People's Daily* publicized He's viewpoints. The State Council's internal publication also summarized the article. These publications signalled that He's reform proposal was on the centre's political agenda. This probably spurred action by the government of Anhui Province to act. Three vice-governors endorsed He's views in a written memo, and his proposals were referred to specialists for examination. However, the corridor of action was clearly delineated by the leadership struggles that had ensued after the June 1989 crisis. Accordingly, write Chen and Chun, 'it was less the practical significance of He's ideas and the deepening of rural reforms which were discussed, but rather if they conformed to the present "regulation and consolidation work" that unfolded country-wide, or if they even contravened it.'[312] Not surprisingly, the experts found that He's ideas did not conform to current political regulations, and his proposal was shelved.[313]

One year later, in early February 1991, He was unexpectedly called to report to the State Council Research Office. There, he was told that Premier Li Peng had read He's second essay and generally endorsed it. However, the Premier also cautioned that such a programme could not be implemented on a large scale, because 'different opinions' to such a reform existed.[314] Obviously, He's proposals had been read by several high-ranking cadres and had probably even been discussed in the Politburo's Standing Committee. Although the Premier endorsed experimentation with grassroots fiscal reforms, he was understandably unwilling to give written proof to that. Other provinces could misunderstand this as an official endorsement and might feel encouraged to stage their own experiments.[315] He continued his research, drew up a detailed argumentation, formulated concrete and manageable measures, and submitted the package to the provincial government.[316] Again, several months of silence ensued. In Summer 1991, however, Long Nian, a vice-governor of Anhui Province in whose responsibility poverty alleviation fell, asked He to try out his policies that same year in poverty-stricken Linxian County. Unfortunately, however, a heavy flood wreaked havoc on the county, and the experiment had to be abandoned.[317]

After Deng's 'southern voyage' in early 1992, it was the leadership of Anhui Province who brought the reform back on the agenda. Again, it was a mixture of chance and personal relations that shaped the next step. At a conference, He Kaiyin had gotten to know the famous rural economist and former Party secretary of Fuyang Prefecture, Du Runsheng. Du introduced He to the Party secretary of Fuyang, who was willing to try out He's idea. Interestingly, tacit approval of the experiment at all superior levels was not enough to ensure its implementation. Obviously, the leading cadres of Yingshan County, which was chosen for the experiment, were evenly split on the issue. While some realized the policy's potential for improving rural stability, others worried that the townships would resist it. He Kaiyin also introduced his policies to the leadership of Guoyang County in the same prefecture. Despite the magistrate's endorsement, however, concrete steps had to be postponed because of the impeding county government elections. Throughout 1992 and early 1993, He Kaiyin continued to lobby his policy in Anhui Province and beyond. Although good intentions abounded, his plans did not reach the implementation stage.[318]

Support and resistance

Apart from natural disasters, what were the factors that determined whether experiments were implemented or abandoned? A comparison of the political processes leading up to the implementation of such reforms in two different provinces will shed light on the complex interplay of personal initiative, organizational resistance, and issue-specific coalition-building. The first experiments were conducted almost concurrently in 1993 in Xinxing Town of Guoyang County, Anhui Province, and in three townships of Zhengding County, Hebei province. One year later, the reform was implemented countywide in Anhui's Taihe County. Although the central government and the local leadership supported all these experiments, those in Anhui faced considerable resistance, while those in Hebei did not.

Xinxing Town

Ironically, the first successful reform experiment was conducted by a town leadership that was probably not aware of He Kaiyin's lobbying. Even more ironically, Xinxing Town was located in Guoyang County, which He had just recently visited. The impetus for the reform experiment in Xinxing Town came from an article that Yang Wenliang, a researcher in the Policy Research Department of the Party Committee of Hebei Province, had published in the *Farmers' Daily* (*Nongmin Ribao*). In an interview, then mayor Li Peijie summed up the leadership's motivation to conduct a fiscal reform experiment:

> It was absolutely necessary to change the existing system! When I became mayor of Xinxing Town in June 1992, the peasant burden was very heavy already. . . . Actually, after sending cadres to every administrative village to investigate, I found out that the problem of village cadres presumptuously raising quotas was extraordinarily serious. Between 1990 and 1992, the average peasant burden in each village ranged between 140 and 190 yuan. At that time, however, the town's average peasant income per year did not even reach 600 yuan! Several cadres stood together and remarked in a casual way that because the funds were a bit tight they would just charge everyone ten yuan more. And so they did. In the eyes of the peasants, the cadres were constantly demanding money, grain and lives [this refers to birth control], and the relationship between cadres and peasants had become very tense.[319]

It becomes clear from Li Peijie's account that there were four main variables which determined the variation in the size of the peasant burden. Three of them have already been identified in the previous chapters. The first was the need to fulfil the tax quotas set by the higher levels. The second was the obligation to achieve development targets. The third was mounting personnel and administrative costs, not the least due to grassroots cadres abusing their power to hire friends or relatives, or to misappropriate public funds. The fourth has so far not been given much attention, but is of equal importance. Since peasants were reluctant to pay their taxes and fees, tax-collecting teams had to be formed. Such teams consisted of hired labour, grassroots cadres and often even included the local leadership. Frequently, forces of the local militia and police had to accompany the tax collectors. According to Li Peijie, the cost of tax collection averaged 10 to 30 per cent of the tax total.[320] Naturally, this sum had to be borne by the peasants as well.

The town leadership believed that a rural fiscal reform had the potential to keep irregular levies in check. This would heighten the peasants' motivation to pay taxes, so tax-collection and personnel costs could be significantly reduced. In addition, a 'rebound' of the peasant burden would be prevented, as grassroots cadres would lose the means to rehire personnel that had been let go. Therefore, reformers calculated the personnel costs, administrative expenses, tax quotas, and costs for social service provision needed for the following year, and divided the total by the acreage contracted to the town's peasants. The result was that the payment of 30 yuan per *mu* land would suffice to finance all these items. Each farmer in Xinxing Town held an average of 1.8 *mu* of contracted land, so the average peasant burden would be reduced from an actual amount of between 140 and 190 yuan to only 54 yuan. Apart from this sum, farmers only had to meet their labour obligation. Land hit by natural disasters and households below the poverty line would be exempt from the levy.[321] The town government signed a three-year tax contract with each household, so the policy became known as the 'tax and fee contracts' (*shuifei chengbao*) system. In contrast to He Kayin's proposal, however, the peasants' had to pay their obligations in cash. Compulsory grain sales to the grain stations continued.

Liu Xingjie and Li Peijie informed the county leadership about the planned experiment. The county leaders were enthusiastic and supplied the reformers with some information about He Kayin's programme. Xinxing's People's Congress officially passed the policy, and a bulletin was drafted to propagate the reform. In addition, the county leadership decided to hold a conference in Xinxing on January 3, 1993, to display their support and to lay the foundation for the experiment's extension across the whole county. At that conference, however, the head of the county PC challenged the reform on the grounds that the new levy exceeded 5 per cent of the previous year's average peasant income and therefore illegally increased peasant burden.[322] This display of opposition was remarkable for two reasons. First, it is very unusual for a PC head to call into question the decision of superior cadres – and to do so publicly. Second, the argument presented was spurious. In 1992, Xinxing Town had collected an average of 22 yuan per capita in agricultural tax and 30 yuan in township and village levies, irregular levies included.[323] The new system would have demanded only two yuan more from each farmer, and irregular exactions would be totally abolished. The argument becomes even more hypocritical considering the fact that for 1993, Guoyang County had demanded the fulfilment of tax and fee quotas amounting to a per-capita burden of 66.3 yuan.[324]

Soon after this attack, the provincial People's Congress joined the chorus and had its mouthpiece, the People's Democratic Herald (*Renmin minzhubao*), condemn the reform as 'lightening the peasant burden in name, but increasing it in reality'.[325] For these reasons, it commanded Guoyang's PC to nullify the document passed by the subordinate legislature. In retrospect, it seems possible that someone in the provincial leadership had opposed such experiments from the beginning. In July 1991, the governor prevented He Kaiyin from attending a national conference on rural problems, obviously because 'someone' had accused him of having fabricated his contacts to the central leadership.[326] As it turned out later, that 'someone' might well have been the governor himself (see below).

As a consequence, the county leadership officially withdrew its backing of the experiment, and two of three other townships that were about to implement the reform abandoned their plans. After an investigation of the prefecture's Legal Affairs Committee, Xinxing's Party secretary was replaced in early March. Two months later, the Standing Committee of Guoyang's PC annulled Xinxing's reform regulation, causing the third township to drop out as well.[327]

Despite the danger of losing his position and having revoked his Party membership, but with the tacit support of the county leadership, Xinxing's new Party secretary Li Peijie covertly carried on with the reform. At the end of 1993 and after two harvesting seasons, Li was able to report some sizable results. First, tax collection after the summer and fall harvests had proceeded much more smoothly than before. In former years, the government employed local militia and police to force recalcitrant villagers to meet their tax dues. In 1993, however, the farmers paid their taxes voluntarily, and collection was finished after just six days even in the most 'difficult' villages. Cadres were personally held responsible for overstepping the law, so irregular fee-taking reportedly was absent that year. A report that was full of praise for Xinxing Town's 'tax and fee contract' system was then sent to the State Council and the government of Anhui province.[328]

Zhengding County

Independent from Anhui, Hebei province had also initiated a fiscal reform programme. At the beginning of 1990, after reading He Kayin's article in a PD special edition, the governor of Hebei Province ordered a feasibility study of the concept's application to Hebei Province.[329] Within three months, aforementioned Yang Wenliang drafted the report. However, the implementation of the reform was, like in Anhui, delayed by an external event. Hebei's governor was transferred to Liaoning Province, and the matter had to be shelved for the time being.[330] In Summer 1992, however, after reading one of Yang's articles in the internal journal of Hebei's Party Committee, the Party leadership gave Yang green light for the reform.[331]

Because of its large agricultural output, Zhengding County was chosen to implement the policy in three townships. The provincial authorities first solicited the opinions of the county leadership, who widely approved of the experiment. In a second step, a survey of 100 peasant households located in ten villages of five townships was conducted. In late May 1993, the provincial Party secretary approved the finalized programme. Finally, the draft reform policy was approved by department-level specialists and academic researchers of the central government.[332] On January 1, 1994, the reform was implemented smoothly. Ten days after the reform had been formally implemented, Hebei's Party secretary decided to extend it to 184 townships in thirty-six counties.[333]

Unlike the experiment in Guoyang, the reform implementation in Zhengding County did not encounter any significant resistance. It is unclear why a vertical alliance of local People's Congress leaderships originated in Anhui and how it developed, but it is almost certain that such an alliance existed. How this alliance formed and why it was able to overpower higher-ranking Party and government leaders cannot be answered here, but is worth more research.[334] One would expect that the provincial People's Congress had the backing of someone important in Beijing, but this seems doubtful given that the experiments in Hebei Province progressed so smoothly. Hebei's pilot reform was the subject of a major year-end conference in Beijing in which the heads and specialists of all relevant central commissions and departments participated.[335] As shown, Xinxing's reform plans were equally well known, so the central government had clearly known of and approved both experiments.

Taihe County

A third occurrence that took place shortly after the two grassroots reform experiments were implemented gives weight to the assumption that the resistance against Anhui's experiments had its origins within the province. In summer 1993, Taihe County of Fuyang Prefecture applied with the provincial government to implement He Kayin's fiscal reform programme. Although it was granted permission to do so,[336] the experiment plunged into a crisis quite similar to that of Xinxing County. On January 25, the government of Anhui province received a short, but sharp rebuttal from the central Ministry of Finance. This document[337] is worth quoting in full:

People's government of Anhui Province:

> It was reported that Taihe County plans to implement a 'reform' of rural taxes and fees in 1994. The agricultural tax, the agricultural and forest special products tax and the township comprehensive fee and the village retention are to be merged and collected in kind as a unified levy based on farmland [holdings]. This kind of practice does not correspond to current tax laws; there are many malpractices [*biduan hen duo*]. Moreover, according to the current national tax collection administrative system, the State Council must decide over changes in the tax system and adjustments of the tax revenue responsibilities. Also, for setting up

> an experimental spot to conduct agricultural tax reforms, a formal report must be submitted to the State Council for examination and approval. Accordingly, we are of the opinion that Taihe's course of action is inadequate and must be corrected.

In May 1994, Anhui governor Fu Xishou ordered the immediate cessation of the Taihe experiment.[338] Despite the sharp wording of the document, Vice-governor Wang Zhaoyao persuaded Party secretary Lu Songjing to override Fu's decision. Wang argued that it would be a loss of face to revoke a permission already given. Lu agreed to let the reform run its course until the fall harvest. Thereafter, its future would be decided. When the time came, the situation took another unexpected turn: when Fu Xishou came back from a trip abroad, he learned at Hefei airport that the centre had decided to demote him with immediate effect. Instead, agricultural specialist Hui Liangyu was appointed the new governor of Anhui province. Hui immediately visited Taihe County to display his support of the reform. On December 18, 1994, Vice-premier and member of the Politburo Jiang Chunyun visited Fuyang with a delegation of other high-ranking central-level cadres. He advised governor Hui Liangyu to extend the experiments, which he did in February 1995.[339]

Judging the available evidence, it is well possible that Fu Xishou was the one who led the resistance against these experiments. The fiscal departments, who at all levels had expressed their scepticism against rural fiscal reforms, might have backed him.[340] Since these experiments were important, but not officially sanctioned, they provided the ideal ground for local-level factional in-fighting. This would also explain why the People's Congresses of Anhui Province and Fuyang's Prefecture were drawn into the conflict. In this respect, the genesis of the RTFR follows the implementation of the Household Responsibility System, the economic liberalization of the early 1980s, and experiments with village elections. The fact that obedience, not disobedience, of orders by the Ministry of Finance was punished supports the hypothesis that someone more powerful than the Minister of Finance wanted these reforms to be carried out. However, the developments surrounding these events suggest another, less fashionable, but compatible hypothesis: despite the obvious importance of these experiments, the top echelon of the leadership was much more occupied with reviving market liberalization, with handling loss-making state enterprises, and, last but not least, with organizing the succession to the ailing Deng Xiaoping. Possibly, the Minister and Vice-premiers into whose area of responsibility the experiments fell followed the developments more closely than the top leadership, who only reacted to emergencies.

After this crisis was solved, local experiments began to mushroom. Fuyang Prefecture alone extended the experiment to five more counties.[341] By 1995, altogether twenty counties in northern Anhui had implemented the reform. Not much is known about how these counties were chosen, but obviously the provincial leadership decided to limit the experiments to the predominantly rural and comparatively poor northern Anhui counties. Here, the problem of forceful burden collection had been especially serious, and violent incidents involving cadres and farmers had become objects of concern even in the central leadership.[342] In addition, Sichuan and Jiangsu also initiated such reform programmes after having sent township-level cadres on a field trip to Xinxing Town. Obviously, other provinces had quickly learned about the experiment despite the secrecy with which it was conducted.[343]

However, there are indicators that not all was as well as the reports made believe. An interview with a former leading cadre who at that time served in the Peasant Burden Section (*nongmin fudan ban*) of the Agriculture Committee of Anhui Province revealed that the Taihe experiment was not as successful as is reported today. In striking contrast to the cheerful leadership of Taihe County, which dubbed 1993 'Increase Peasant Revenue, Reduce the Peasant Burden Year',[344] the interviewee points out that

> the aim of the reform was not the reduction of burden. There were three aims, and burden reduction was not one of them. First, Taihe wanted to simplify and reduce the costs of revenue collection. Second, it wanted to increase collection of food grains. Third, it wanted to increase its revenue.[345]

In 1997, Fuyang city applied the system to all its counties, but with two important modifications: first, taxation in cash was reintroduced; second, Fuyang reverted to combining per-capita and per-acreage taxation. With a whopping 100 yuan per person and 30 yuan per *mu* levied, the peasant burden almost doubled to 140 yuan per capita, at least in Taihe.[346] The interviewee confirms that burden was not reduced, but steadily increased in the course of the reform. The problem, he said, was that

> two models existed. The provincial government had sanctioned one set of policies, but the actual regulations guiding implementation of the second reform round differed from that model. One model received approval, but Fuyang implemented another. The difference was that the Fuyang model brought more burden to the peasants than the one sanctioned by the province. In fact, the reform added 1 billion yuan to Fuyang's revenue.[347]

Similarly, reform implementation in Huaiyuan County, had worked quite well in the first round, 'but was not able to solve the problem of illegal levies and apportionments'.[348]

Official recognition

Despite these problems, the initial reform experiments were considered a great success by the provincial leadership, and He Kaiyin's model of collapsing the agricultural tax, the agricultural special products tax and the *santi wutong* into one unified tax became very popular. First of all, the costs of tax collection were reduced significantly. Collection was now restricted to the harvest periods and conducted openly, and the majority of the peasants were reported to have widely accepted the new system. They would deliver their grain, get a receipt, and all other attempts to collect money would be illegal. As a consequence, tax collection teams visiting each peasant household had become not only redundant, but also illegal. Second, the new system was considered fairer, because land holdings became the sole criterion on which taxation was based. Before the reform, the amount of land contracted determined agricultural tax obligations, while the *santi wutong* were calculated on a per capita basis. Therefore, large households with little land would be charged considerably less, and small households with large landholding considerably more than before.

The reform in Xinxing Town was reported to have reduced the peasant burden by more than one third, and to have motivated peasants to invest in agriculture. In addition, the reduced tax collection costs allowed Xinxing Town to get rid of more than 300 village cadres, which further reduced the peasant burden.[349] Taihe reported similar successes. According to Chen Guidi and Wu Chuntao, tax collection was finished within a record five days, and only one peasant household in the whole county did not pay its taxes. While 500 persons had registered burden-related complaints in 1993, the reform reduced the number to zero. Most important, the peasant burden was reduced 'more than half'.[350] In addition, the year 1994 registered record harvests, for which the positive effect of the reform on the peasant's motivation to engage in agriculture was made responsible. Finally, the reform had a very positive effect on public finances. Taihe County had made cost and revenue projections on the base of revenue estimates that were based on average yields and pre-reform grain prices. Based on these estimations, tax revenue was projected to reach 70.3 million yuan.[351] Due to the record harvest and a positive price development, however, the 81,929 tons of wheat and corn collected bestowed more than 94.7 million yuan in revenue on the county government.

Despite continued endorsement by central-level cadres of vice-ministerial rank, it was only in 1996 that local-level reform experiments were for the first time officially recognized and explicitly approved. On December 30 that year, the CPC Central Committee and the State Council jointly passed the 'Decision to Work Well in Alleviating Peasants' Financial Burden', which also became known as 'Document 13'.[352] This was also the first time that the public peasant burden agenda and the hidden rural fiscal reform agenda became connected. However, the document makes clear that the two agendas were far from compatible. Specifically, it continued the worn strategy of reducing burden by limiting the fiscal powers that the localities enjoyed:

> During the Ninth Five Year Plan Period, no new categories of agricultural tax will be introduced and the agricultural tax rates set by the government

> will not be increased. Local authorities do not have the right to set up tax categories or increase tax rates. Illegally created tax categories and illegally increased tax rates should be cancelled. Taxes on special agricultural products and on slaughtering animals should be levied according to facts and no quota should be set on each peasant or on a village according to head count or acreage.

This, however, was exactly what the reform experiments in Anhui had done. The agricultural tax was increased to include the *santi wutong*. In addition, the ASPT was almost never levied 'according to facts', since establishing these 'facts' had proven extremely impractical and expensive.[353] However, at the very end, the document finds that

> Alleviating the peasant burden is a long-term task. We must intensify reform efforts and solve the problem of excessive burden once and for all. *Experimentation with fee-sharing schemes in some localities and pilot reforms in tax and fee collection in some key grain producing regions may be continued.*[354]

Hence, the document stresses the need to stick to the conventional burden reduction approaches and does not encourage new experiments. At the same time, however, the existing experiments were given a favourable opinion. In his speech at the National Rural Work Conference on January 13, 1997, Vice-premier Wen Jiabao went one cautious step further and signalled that a rural fiscal reform had indeed become a national policy option, but that the centre was not yet ready to take the risk:

> For several years, some key grain producing regions, most notably fifty counties in seven provinces like Anhui and Hebei, have conducted experimental rural tax and fee reforms within certain limits. They have achieved notable results and have accumulated much profitable experience. The centre is of the opinion that experiments with this reform may be continued, but they should *not yet* be universally extended. This stance rests on two deliberations: the reform touches upon several deeply rooted problems in the [political] system, and it takes us on a path leading towards far-reaching reforms. If we continue the reform, we must consider in a circumspect way the reform of the current grain purchase system and the fiscal system with the agricultural tax at its heart. This is a far-reaching and complicated matter that has to be comprehensively engineered and planned. The work in the trial areas must depend on leadership, and those places that have received permission to conduct experiments should earnestly conduct valuable experiments, and they should summarize their experiences.[355]

Seen through our lens of intergovernmental relations, the still prevalent approach to burden reduction was the recentralization of local rights and responsibilities. The fiscal reform experiments, however, depended on the decentralization of decision-making authority. Given the support of the central leadership and the conflicting nature of the two policies, one would have expected less recentralization and more decentralization to occur. This, however, was not the case. Nor were attempts made to combine the two policies, for example by clearly delimiting a sphere of autonomy in local decision-making. Although grassroots fiscal reforms were now an official policy, the implementation of the prevailing recentralization policy was continuously pushed forward. As the next chapter shows, this very soon led to serious conflicts that threatened the survival of the reform experiments.

Conclusion

This chapter has shown that the peasant burden was on the political agenda consistently, but with varying intensity. This intensity often coincided with the intensity of burden-related reporting in the national print media, and there is strong evidence that the central government was responsible for this. In other words, the public agenda did not determine the political agenda, but the other way round. Second, this discourse gravely neglected the complexity of the peasant burden problem. The previous chapter has shown that the concept is highly misleading, as it suggests that all or at least the majority of the peasants suffered from excessive exactions, a claim that can be clearly refuted. No doubt, the leadership is well aware of the heterogeneity of the peasant strata. Nevertheless, peasants were addressed as one identity group whose lot the Communist Party pledged to defend against greedy local officials.[356] In effect, the centre formed a vocal alliance with the peasant strata, thereby increasing its legitimacy at the cost of local officials. In line with this discourse, central documents aimed to reduce burden by circumscribing their powers, most notably by imposing hard budget constraints while simultaneously reducing their revenue sources. Interestingly, rural fiscal reforms entered the political agenda long before the public peasant burden discourse reached its first peak in 1993. This peak would have been a good opportunity to connect these two issues, but they remained separated. This might have to do with the fact that the prime aim of He Kaiyin's reform proposal was not the reduction of the peasant burden, but the reinvigoration of the agricultural sector. Burden reduction was an important means, but by no way the aim. Of equal interest is how the reform proposal entered the agenda, and how it developed. Personal relations were exceedingly important and external factors such as natural disasters, political upheavals or simply leadership changes played a great role in determining the placement of the issue on the centre's agenda. In fact, they played a far greater role than burden-related demonstrations and riots that allegedly threatened China's political stability. Never did the reformers use these incidences to argue for their agenda, nor did the central leadership's interest in the reforms coincide with the peaks in peasant burden reporting. Finally, insight has been gained regarding struggles of how various (collective) actors in the process fought to realize rural reforms. A reformer's alliance had begun to form very early, consisting mainly of the leaders of the executives at all concerned levels. However, resistance equally existed, and such resistance tended to be displayed vertically, i.e. along functional lines, rather than horizontally, i.e. along territorial lines. Finally, while all relevant Party and government organizations in Anhui Province made the rural fiscal reforms a priority issue, the same cannot be said of the central leadership. Of course, the experiments had the support of the Premier, and the number of relevant supporters grew as the reforms unfolded. Their support was strong enough to replace a governor who had followed the Ministry of Finance's order to halt the Taihe experiment, but they proved either unwilling or unable to halt the progress of other policies that threatened the experiments' survival. For example, the tax division system was implemented in the same year as the Taihe experiment, and the centre did not use the newfound riches to support those reforms. On the contrary, despite tighter finances, further target achievement programmes were handed down that very same year. Regardless of these obstacles, however, the reform eventually entered the centre's official agenda, if only juxtaposed to conflicting burden-reduction programmes.

5 From innovation networks to competition under hierarchy

This chapter analyses the process of how the RTFR became a national policy. It identifies the central- and local-level actors that were involved in designing the RTFR, and explores how they were related to each other and to the main actors of the previous phase. Furthermore, it analyses the instruments that these actors used in influencing the formulation of the policy, and relates both the constellation of the actors and the instruments they employed to the resulting policy. The main point to be made is that a crucial juncture was reached at this point of the reform process. While the agenda-setting stage had been dominated by decentralized policy networks and bottom-up innovation, the centre now took over the reins, discontinued this steering mode, and once again stimulated competition under hierarchy to come up with a policy that was both unified and adapted to local circumstances. Although the dynamics of the reform process clearly shifted from agency-heavy to institution-heavy, it will be seen that agency still played a crucial role at this stage. The chapter starts by outlining the different models that had emerged in the phase of local-level experimentation, and goes on to show how implementing a rural tax reform quite suddenly became a priority issue for the central government, and how networked bottom-up innovation was instantaneously replaced by competition under hierarchy. The subsequent analysis of the policy document that formed the legal base of the reform, serves to reiterate this point.

Different approaches to rural fiscal reform

During the phase of networked bottom-up innovation, a number of approaches for reforming the rural fiscal system were tried out. These approaches, which can be grouped into three basic models, had much in common. First of all, since local experimentation cannot tackle systemic problems for which central-level policies are responsible, they were restricted to the townships and villages. None of them encompassed the county level, and they did not target the fiscal flows between counties and townships. Strictly speaking, these experiments did not (and could not) change the structure of intergovernmental fiscal relations, but attempted to regularize tax collection at the grassroots levels, and to render it both more efficient and transparent. In addition, all models included some form of a 'contract' between the government and the peasants which clearly stated the rights and responsibilities of all involved parties. This is a clear hallmark of negotiation as a policy instrument. Finally, most of them specified how the exactions would be used. In other words, cautious attempts were made to formulate a budget of grassroots revenues and expenditures.

The first model is that proposed by He Kaiyin, and it entailed the most far-reaching changes. It was also the most popular one, as it was implemented in all of northern Anhui. The second model merely transformed the township comprehensive fee and the village retention into a new tax category. The third left the system largely unchanged, but made it more transparent (Table 5.1).

As described above, the Taihe model linked fiscal exactions and mandatory grain purchases. The agricultural taxes were merged with the township and village levies and collected in kind, and this revenue was used to meet the grain purchase quotas prescribed from above. In Anhui's Taihe County, for example, each person had to submit 50 kg of grain and maize each year for each *mu* of contracted land. This amount was not to change for three years. Instead of buying grain from the peasants, the grain purchase stations now obtained their grain from the government. If peasants were not able to pay in kind, they could pay in cash. Poor households profited from partial or full exemptions, and peasants were given the right to refuse excessive exactions. Their obligations and payments were registered on a 'peasant burden supervisory card' (*nongmin fudan jiandu ka*). The tax grain was jointly administered by the grain stations and the fiscal departments. The former converted the surplus grain to cash and submitted it to the township fiscal department, who could now meet the tax quotas imposed from above. The remaining sum was used to meet legitimate township and village expenditures. In contrast to the pre-reform practice, however, the villages lost considerable autonomy in generating and allocating revenue. Village expenditures were now administered by the townships (*cun you, xiang guan*).

The 'Tax for fee' or 'Rural Public Welfare Facilities Construction Tax' model was far simpler, as it merely converted the township and village levies into a tax. According to a report on the implementation of a Tax-for-fee reform experiment in Hunan's Wugang County, this model was drafted in the Ministry of Finance. The former head of the RTFR small group in Anhui's Wuhe County, who was among the first to implement this model, confirmed this.[357] In the Tax-for-fee system, the farmers still had to pay their dues in cash. As had been the case with the *santi wutong*, the new tax could not exceed 5 per cent of peasant average income and was levied at a per-capita basis. Average peasant income was assessed at the village instead of the township level, and peasants holding no land had to pay only 70 per cent of the new tax. The agricultural taxes, however, continued to be managed

Table 5.1 The three basic models of rural fiscal reform in China

1 **Collection of agricultural taxes and fees in kind** (Taihe County, Anhui Province)

	Santi wutong			
Agricultural taxes	*Township comprehensive fee*	*Village retention*	*Other fees, e.g. education*	*Grain purchase quotas*
Belong to fiscal department	Belongs to township	Township manages, village uses	Rescinded Later: only 'legal' fees to be collected	Abandoned at township level and below Grain accounts settled by township-level fiscal department and grain department Grain department pays fiscal department in cash
Agricultural taxes and *santi wutong* merged and collected together Collection not in cash, but in kind. Exceptions possible No change in collected amount for three years Comprehensive levy: calculated per <u>*mu*</u> land (50 kg of food grains in Taihe) Exemptions for poor households				

Farmers have the right to refuse further exactions not sanctioned by law 'Burden card' system to make extractions transparent

2 **'Tax for fee'** (Wugang City, Hunan Province; Wuhe County, Anhui Province)

Agricultural taxes	*Santi wutong*	*Other fees, e.g. education*	*Grain purchase quotas*
No change	Converted to 'Rural Public Welfare Facilities Construction Tax collected in cash Tax rate less than 5% of peasant average income Reductions for landless (30%) and abandoned land (70%) Figures calculated at village level	No change	No change

Farmers have the right to refuse further exactions not sanctioned by law

3 **Land Comprehensive Burden Responsibility System** (Yangdang Town, Zaoyang City, Hubei Province)

No change in system, only specification of items for which taxes and fees are collected, as well as how much is to be collected (income percentages for taxes, concrete amount for fees).
Figures calculated at village level.
Maximum burden not to exceed predefined ceiling (1997: 250 yuan; 1998: 220 yuan per *mu* land)

Source: Based on X. Ma, H. Huang and C. Jiang, *Wo guo nongcun shuifei gaige yanjiu*, Beijing: Zhongguo jihua chubanshe, 2002, and L. Wu, 'Nongcun shuifei gaige shidian jige moshi de bijiao', *Jingji luntan* 17, 2002.

separately. Thus, the new model did not make as much of a difference for the farmers as it did for the townships and villages, but it is easy to see why it was the Ministry of Finance's favourite: existing laws and regulations were not violated. Similar to He Kaiyin's model, the new tax was collected and administered by the townships, which were therefore given control over revenue that had formerly belonged to the villages.

The third model was the least far-reaching one. The old system was largely retained, but a ceiling was set on the peasant burden by clearly specifying which levies were legitimate and how much could be collected. However, this ceiling could be easily changed, and new items were quickly added to the catalogue of legitimate fees.[358] As in the first model, peasants were granted the right to refuse exactions not listed in the contract.

The other counties applied all possible combinations of these models. Meitan County in Guizhou Province, for example, combined features of the first and second model. As in the first model, the agricultural tax and the town and village levies were extracted together and in kind. However, the township comprehensive fee continued to be calculated on a per capita basis, as in the second model, and it was administered separately from the agricultural tax. The village levy, on the other hand, was calculated on the basis of land holdings and integrated into the agricultural tax as a surcharge. Different accounts were set up for each kind of revenue, and village finances were administered by the townships.

Policy formulation

Soon after rural fiscal reforms had been formally put on the official agenda, the policy process faced a sharp disruption. At first, the issue took a back seat at the centre, and reform experiments were allowed to continue at the local level. The main reason for this inactivity is that after the death of Deng Xiaoping in February 1997 and with the Communist Party's fifteenth National Congress approaching, central leaders had more important matters to attend to. After the new leadership had been sworn in before the National People's Congress in March 1998, however, the issue quickly received high priority. It is interesting to examine in detail how the policy formulation process commenced, because it illustrates the crucial role of agency in the central government. The bottom-up reformers were almost completely bypassed in this process, and personal considerations of high-level leaders took precedence to crafting a sustainable rural tax and fee reform.

One sign marking the elevation of rural fiscal reforms to a higher priority on the central agenda was the position of the speakers that began to speak out for it. Whereas central-level politicians of vice-ministerial rank or below had publicly dealt with the issue before, commitment was now signalled from the top leadership. Just after having been appointed Premier, Zhu Rongji stated at a press conference that he would strive to achieve five reforms during his tenure, the first of which was the reform of the grain distribution structure, another being 'the further improvement of the tax system'.[359] A commentary by the Xinhua News Agency four months later further clarified these issues. 'Reform of the grain distribution structure' meant buying more grain from farmers at state prices, which in the meantime had exceeded market prices and therefore amounted to subsidies. As for 'improving' the taxation system, the Tax-for-fee reforms were explicitly mentioned. Clearly, a model linking agricultural tax collection with the state grain purchase system was not considered.[360] Hence, although local-level reformers overwhelmingly favoured He Kaiyin's model (which had been tested in a number of localities), the new Premier rather embraced the less far-reaching Tax-for-fee reforms. Shortly after, Chen Jihua, the new head of the State Planning Commission, reported to the NPC that '[w]e must gradually phase out transitional fees collected to make up deficiencies in administrative expenses in line with the principle of 'more emphasis on taxes and less on fees'.[361] At the State Council's first meeting on March 24, 1998, Zhu became more concrete and explicitly stated that Tax-for-fees was the preferred model:

> Along with the deepening of the [economic] reforms and economic development, several new problems need to be solved. A rather prominent problem is that wanton fee-taking [*luan shoufei*] is getting worse and worse in the cities and the countryside alike. Property is taken by force, and a lot of money is embezzled. For the enterprises, this is hard to bear, and the population is extremely dissatisfied. [To solve these problems], [t]he method of reform is the implementation of 'Tax for fees'. It will not be allowed anymore to levy those fees that should not be levied. As for the legitimate fees, we must strive to change them into taxes and include them into the regular budget after careful examination. For those fees that need to be levied, two separate tracks must be established for revenue and expenditures. The Ministry of Finance and the General Tax Office must take the reorganization of taxes and fees very seriously.[362]

The end of local experiments

Shortly after, the passage of the Regulation for the Purchase of Grain (*Liangshi shougou tiaoli*) on June 6, which resulted from Zhu's determination to reform the grain distribution system, sounded the death knell for most of the promising fiscal reform experiments that had been conducted in Anhui Province and elsewhere. Very similar to Document 13, which was discussed in the previous chapter, the Regulation sought to reduce the peasant burden by diminishing local political autonomy. In order to prevent local governments from levying irregular fees while purchasing grain, Article 11 of the Regulation decreed that

> state-owned grain purchase and storage facilities shall, in accordance with existing state regulations, collect agricultural taxes from the peasants on

> behalf of the government. They shall not deduct any other taxes or fees on behalf of any department or individual. Township (town) or village cadres shall not collect overall planning fees, accumulation funds, or any other taxes or fees at the grain stations. No department or individual shall entrust state-owned grain purchase and storage facilities with the deduction of any taxes or fees besides the agricultural tax.

Clearly, these rules rendered the reform experiments that had combined tax and fee collection with compulsory grain purchase explicitly illegal. Agricultural taxes had to be collected separately from levies and fees, and only the agricultural tax could be collected in kind.

Given that a Regulation could, due to its higher legal standing, not be disregarded or circumvented like an Opinion or a Notification, the Regulation spelled the end of experiments with He Kaiyin's model. As opposed to the December 1996 circular, the Regulation did not grant any exceptions to ongoing experiments. As a consequence, all experiments but one were halted. Fuyang prefecture continued its reform experiments, but in order to conform to the Regulation, the agricultural tax was now levied in cash and per capita,[363] rendering the original reform design meaningless.

The centre takes over

September 25, 1998, was a remarkable date for China's rural tax and fee reforms, because within just one day, not only was the issue moved to highest priority at the central government level, but also was the decision made to follow He Kaiyin's model instead of the Tax-for-fee method initially favoured by Zhu Rongji. The reason for this sudden turn in events was that on this day, Party secretary Jiang Zemin for the first time mentioned rural fiscal reforms, exhorting his cabinet to 'reform and regularize the rural tax and fee system, and explore policies to get at the root of the burden problem'.[364] This statement, made during a study tour through Anhui province, not only signalled an elevation of the issue, but also Jiang's support of Fuyang's experiments.

That same month, a Rural Tax and Fee Reform Working Group (*nongcun shuifei gaige gongzuo xiaozu*) was established by the State Council's Leading Small Group for Agricultural Work (*nongye gongzuo lingdao xiaozu*). The Working Group was to be led by the Ministry of Finance, but the Ministry of Agriculture and the State Council's Financial and Economic Leading Small Group were to participate as well. At the time of its establishment, the group only consisted of the heads of the three units,[365] so it was nicknamed 'group of three persons' (*san ren xiaozu*). The theme of the Third Plenum of the Fifteenth Party Congress, which met between October 12 and 14, also testifies to the relevance rural issues had suddenly taken. Most notably, the Plenum passed the Decision Regarding Several Major Problems in Agricultural and Rural Work.[366] Already at this Plenum, the concept of a

New Socialist Countryside, which became the guideline for rural development in early 2006, was born, and the Decision formulated a timetable for its establishment until 2010. The Rural Tax and Fee Reform thus an integral part of this programme,[367] and the 'relevant departments' of the Party's Central Committee were empowered to swiftly draft a Rural Tax and Fee Reform programme.[368] On October 27, the Group of Three wrote Zhu Rongji a letter to suggest the nationwide extension of the trial reforms within two years. Given that rural fiscal reforms had lain dormant at the central level for so long, it is astonishing that Zhu now could not wait to see them implemented. He replied that

> implementation does not need to be delayed until 2000. Let us first pass a document to the effect that each province and city can, considering their specific context, freely decide on the timing of the reform. Several provinces should make the first step already next year.[369]

According to Chen Guidi and Chun Tao, Zhu personally reiterated his stance to Xiang, Chen and Duan few days later:

> Discussions with several provincial and city leaders have revealed that this reform is already ripe [for implementation] and should not be delayed any longer. Of course, we must work thoroughly, but the centre does not have to regulate all the details and standardize the time of implementation. In fact, some provinces have already implemented it in several regions. The more Leading Small Groups and offices there are, the more difficult it becomes to achieve something. The State Council has already authorized you to convene all the departments you need for consultation on and handling of this issue.[370]

Zhu's remarkable change of mind regarding the reform model he favoured occurred at about the same time, and is summed up nicely in a statement he made to the *Guangxi Ribao*:

> A few years ago, I already had a good idea, and I have thought about it for many years. My idea is to integrate all legitimate fees into the agricultural tax and let the village cadres 'eat the imperial grain'. It suffices to raise the tax a little bit, the peasants can afford it. Besides the agricultural tax, all other fees must be considered illegal. Apart from the tax departments, no other units or individuals have the right to demand fees from the peasants. Those who do, break the law. In this way, those who engage in illegal levying have no more excuses, and the farmers can be sure that justice is on their side when they refuse to pay illegal fees. It is totally feasible to solve this problem if the centre and the localities share the same ideas and the same understanding. This [reform] is beneficial for the peasants, beneficial for the state, and it helps to quell

> illegal fees, to improve the relationship between cadres and the population, and to put an end to corruption. Hebei has been a trial province for several years, and practice proves that the reform is good.[371]

Another month later, a State Council notice formally entrusted the Group of Three with the 'responsibility to draft a Rural Tax and Fee Reform programme' and urged it to 'produce such a programme as fast as possible'.[372] Finally, Fuyang's experiments were rehabilitated. At the National Fiscal Work Conference held in Beijing in December 1998, Zhu remarked that 'the reforms in Anhui and Hebei have been conducted in a good fashion, and they should continue'.[373]

On March 5, 1999, in his Report on the Work of the Government, Premier Zhu Rongji stressed the need for rural fiscal reforms twice. First, he reiterated the general need to 'proceed with restructuring the financial and tax systems, giving priority to comprehensively reviewing and standardizing all collectible fees and gradually replacing these fees with taxes'.[374] This policy should not be restricted to the rural sector, where it had been carried out on a trial basis for already five years, but should be a cornerstone of a national tax reform.[375] Second, Zhu specifically urged that '[w]e should lose no time in formulating and implementing plans for replacing rural charges with taxes so as to protect the peasants' legitimate rights and interests and to completely resolve the peasant burden problem'.[376]

At about the same time Zhu delivered his work report, Anhui's Ministry of Finance produced an evaluation of the various fiscal reform experiments conducted so far. As could be expected, its suggestions bore the hallmarks of He Kaiyin's reform model. Specifically, it proposed fusing the township and village levies into an increased agricultural tax, whereby part of the village retention was to be levied as a surcharge. A novelty was the proposition to establish a 'public reserve funds'. All villagers should decide by majority vote on village-level public projects that could not be financed by means of budgetary revenues. Each new project had to be voted on, a decision-making process that was named 'one undertaking, one decision' (*yi shi yi yi*). Each villager would then have to pay the specified sum into the fund, which was to be administered by the village administration.[377]

That same month, Xiang, Chen, and Duan submitted to the State Council an Investigation and Research Report Concerning Major Policy Problems Related to the Rural Tax and Fee Reforms (*guanyu nongcun shuifei gaige youguan zhongda zhengce wenti de diaoyan baogao*). This report only summed up the reform experiments and put forward some general ideas, but concrete policy suggestions followed shortly after. At the fiftieth Bureau Meeting of the Prime Minister (*zongli bangong huiyi*), the Working Group submitted a report concerning the general principles of the reform, and at the subsequent fifty-fifth and fifty-seventh Bureau meetings, details like tax rates and the geographical scope of the trial reform were clarified. The general principles of the report were handily summarized in a linguistic trio. The overall aims were the 'reduction' (*jianjing*) of farmers' burdens, the 'standardization' (*guifan*) of the collection and administration of taxes, fees and levies, and the 'stabilization' (*wending*) of tax and fee levels. The methods by which these aims were to be achieved were remarkably similar to those that He Kaiyin had originally proposed, and which had also been put forward in the Report of Anhui's Ministry of Finance.

They were also packed into easily memorizable phrases: 'three scrappings, two adjustments, one reform' (*san quxiao, liang tiaozheng, yi gaige*). Along with the township comprehensive fee and all irregular fees, funds, allotments and fines, the slaughter tax was to be rescinded. The latter had been a source of peasant discontent, as it was levied on a per capita basis regardless of the number of animals actually slaughtered by a person. Adjustments were foreseen for the agricultural and agricultural special products taxes. The Group of Three suggested setting the agricultural tax rate at 7 per cent of a locality's average harvest between 1993 and 1998. The 'one reform' referred to collecting the village retention as a surcharge to the agricultural tax, instead of levying it separately. This surcharge, which was not to exceed 20 per cent of the agricultural tax rate, was to be 'administered by the township and used by the village' (*xiang guan cun yong*).[378] This report, which was based on the research report delivered by Anhui's Ministry of Finance, served as the blueprint for the central-level document inaugurating the reform one year later.[379]

In May, four counties of Anhui province[380] were chosen by the provincial government as reform pilots.[381] While the reform experiments proceeded in those four counties, the State Council formally submitted the policy draft to the Politburo and its Standing Committee for deliberation. On February 3 and 18, 2000, both bodies received final reports detailing the experiments in the four counties. The Politburo approved the reports, and on March 2, 2000, the Notification by Communist Party's Central Committee and State Council Concerning the Promotion of Work in Rural Tax and Fee Reform Trial Areas, often referred to as 'Document No. 7' (*qi hao wenjian*), was officially passed.[382]

Competition under hierarchy: refining the Rural Tax and Fee Reform

Document No. 7 must be considered the legal base of the RTFR, although a large number of policy circulars were passed later to fine-tune the reform (see next chapter). Despite the importance of the reform, which was dubbed 'the third revolution in the countryside', none of these documents obtained the status of a Regulation, let alone a Law. This signifies that the reform was considered 'work in progress'. Hence, Document No. 7 is rather short. It consists of five sub-headings. The first justifies the reform, the second outlines its general principles, the third regulates implementation, the fourth lists some 'complementary measures' (*peitao cuoshi*), and the fifth urges leading cadres to take the reform seriously. Remarkably, the justification of the reform is almost as long as the rules guiding its implementation. Against the backdrop of the agenda-setting process outlined in the previous chapter, this narrative is of crucial interest. As was explained, the prevailing burden-reduction policies made local leaders responsible for the 'burden problem' and therefore sought to reduce their decision-making powers. The success of the rural fiscal reform experiments, on the other hand, depended on local initiative. What course did Document No. 7 take? Simply, it continued the narrative of the centre and the localities of two competing forces. As the following paragraphs will make clear, the central government portrayed itself as a pro-peasant reform force and made no mention of the fact that the reform had in fact originated at the local level. Yet at the same time, the reform document was not much more than a blueprint of what had been achieved at the local levels, lacking a comprehensive approach to tackle the systemic deficiencies outlined in Chapters 2 and 3. Finally, it will also become clear that the centre once again stimulated competition under hierarchy to fine-tune a reform whose purpose basically was to remedy the negative effects caused by competition under hierarchy.

Narrative

The document starts off by lauding the Party's achievements in improving the peasants' lot, 'especially after the Reform and Opening period'. The Household Responsibility System, the price and purchase policies for agricultural products, the 'improvement of rural distributive relations' and a spate of burden-reduction policies are given as proof. 'However', the document continues, 'the current rural tax and fee system and exaction methods are not yet completely reasonable. The peasant burden is heavy, and the problem of non-standard tax and fee collection still exists'. The following lines are worth to be quoted in full, because they not only illustrate the kinds of undesired burdens the farmers are facing, but also make clear who, according to the centre, is to blame for this:

> There are a several places and departments that do not heed the centre's repeated orders and grab for the farmers' pockets as they please. Various kinds of fees, funds, fines and apportionments for various projects face the farmers in great numbers. There are several localities that misreport the farmers' income, collect the village retention and the township comprehensive fee over the quota, and force farmers to substitute cash for labour. There are several localities that violate state regulations by collecting the special products tax and the slaughter tax on a per capita base or according to a household's land holding. There are several departments that demand the grassroots level to engage in target achievement activities (*dabiao shengji*) which had often been forbidden, but were never stopped, and the expenses were eventally apportioned among the farmers. There are several grassroots cadres

> that use illegal measures to forcefully take money and goods from the farmers, nurturing ugly occurrences and public unrest. These problems severely infringe upon farmers' material benefits and democratic rights, frustrate their productive activity, frustrate their feeling towards Party and government and affect social stability in the villages. These problems must be solved from the roots by ways of deepening reform.

Hence, systemic deficiencies are mentioned, but as will be shown below, they are portrayed as local in character and structural in nature. The problematic incentive structure inherent in the current institutional set-up regulating central–local relations does not seem to have found its way into the policy formulation process.

In addition, macroeconomic reasons are provided to justify the reform. Peasant income, the document states, had decreased because of an oversupply of agricultural products and falling prices. At the same time, TVEs had been developing slowly, their profits had been decreasing, and rural–urban migration had become harder. Again, no connection is being made between these developments and government policy or political structures. Rather, these changes are being portrayed in a way quite reminiscent of the Marxist logic that the development of the economic basis is historically determined. However, the document departs from the classic logic when it suggests that this basis can be changed politically:

> Under these circumstances, the key to increasing peasant income is to suitably react to the demands posed by [this] new stage in agricultural development. In order to open up new roads and new territories for increasing peasant income, agricultural and rural economic structures need to be strategically adjusted. This, however, is a rather extended process.

The document identifies rural development as the bottleneck of national economic strength, and demands that 'in the long term, the distribution of national income must tilt towards the farmers, and efforts to decrease their burden must be given prominence to'. Only in this way could peasants be motivated to 'increase their investments and improve production', and 'only in this way can the purchasing power of the villages be strengthened'. Another plus of the reform was that 'rural cadres are released from extracting grain and cash', which would not only improve cadre–peasant relations and further the rule of law, but also contribute to the 'transformation of grassroots government and the simplification of administrative structures'.

It is stressed that implementing the RTFR would be a lengthy and complicated process, because various interests were affected, and because the problems were historically determined. Specifically, 'the simplification of township-level administration structures, personnel reduction, and the readjustment expenditure structures . . . cannot be achieved in a single step'. In order to solve this problem, a double strategy of 'not only relying on a unified national policy, but also fully incorporating every locality's unique circumstances and practicing dispersed policy-making' was to be achieved, which amounts to a clear commitment to competition under hierarchy for fine-tuning and implementing the reform.

The second part of the document lists the 'guiding thoughts' and 'basic principles' of the reform and thereby outlines the main features of the 'unified national policy'. The 'guiding thoughts' can be easily summarized: reduce the peasant burden, and at the same time improve local governance. The 'basic principles' state how this should be done: the peasant burden must be reduced by setting up a 'scientific standard' system for the allocation of public goods, and implementing a 'convenient and practicable' tax collection system. At the same time, it to be ensured that village- and township-level governments continue to function 'smoothly'. How, one might ask, can this be achieved in the face of a significant reduction of revenue? The answer lies in the 'complementary measures' to be undertaken by the local governments, such as 'transforming government functions' (*zhuanbian zhengfu zhineng*), trimming township-level governments, reducing the number of government employees, improving township- and county-level financial administration, balancing revenue and expenditures, adjusting expenditure structures and reducing government expenditures. As can be seen, the RTFR contained regulations to reduce burden as well as to streamline local governments. The 'main reform' (*zhuti gaige*) aimed at reducing farmers' financial burden by standardizing fiscal exaction, while several 'complementary reforms' (*peitao gaige*)[383] were to ensure the long-term sustainability of burden reduction. The following sections describe these two components in more detail.

The Main Reform

The contents of the Main Reform are almost identical to the Opinions submitted by the Group of Three, as the following list shows:

1 Abolition of the various fees, funds and allotments, including the township comprehensive fee and the village education fund.
2 Abolition of the slaughter tax.
3 Gradual abolition of compulsory labour.
4 Increasing the agricultural tax to a maximum of 7 per cent.
5 Standardization of the special products tax.
6 Inclusion of the village retention into the agricultural and special products taxes as a surcharge of a maximum of 20 per cent of the root tax rates.
7 Democratic approval of village-level projects directly financed by the peasants.[384]

As can be seen, there are two new features in Document No. 7. First, compulsory labour was to be phased out within three years. This is a novelty which had neither been part of He Kaiyin's model, nor of the actual experiments.

Second, the 'public reserve funds' proposed by Anhui's Ministry of Finance found its way into the reform, and a ceiling on such funds was specified: the maximum sum levied in the course of such 'one undertaking, one decision' (*yi shi yi yi*) projects was fixed at 15 yuan per person and year.

In addition, the document provided further details regarding the other agricultural taxes. It decreed that the animal husbandry tax was to be lowered to a level below that of the agricultural tax. The agricultural special products tax, another problematic tax, was to be standardized as well.[385] Furthermore, the taxes could be levied in cash or in kind, and the grain stations were empowered to deduct the taxes when buying grain from the peasants. Most important, the taxes were levied on the basis of the land each household had contracted during the second round of land distribution in 1987. Peasants farming other land had to be taxed according to their actual land holdings. Villagers depending on non-agricultural income could be taxed at a level not exceeding that of the new tax, and the concrete amount had to be decided by the Villagers' Assembly. All other exactions were illegal, except in the case of natural disasters.

The designers of the reforms probably foresaw that the implementation of the Main Reform had the potential to plunge grassroots governments into severe financial difficulties, which would further endanger the provision of local services. Therefore, the Complementary Reforms were drawn up to address these issues.

The Complementary Reforms

The slogan underlying the complementary policies was that local government at the county level and below should change its role from 'managing' or 'administering' (*guanli*) to 'serving' (*fuwu*) its constituency, and their aim was making it more effective, more transparent and less costly. The concrete aims were the following:

1 Reduction of territorial entities by merging villages, townships and bypassing prefecture- and township-level governments.
2 Reduction of the number of offices and employees in counties, townships, villages.
3 Strict regulation for the process of hiring, evaluating, remunerating and promoting local government employees.
4 Outsourcing of government services.
5 Strict limitation of local-level fiscal capacities.
6 Separation of government and business.
7 Creation of a supervisory system.[386]

Abstract principles proffered by Deng Xiaoping during the initial stages of the Reform and Opening period, such as 'separating the Party from the state' (*dangzheng fenkai*), gained renewed currency. It was thus held that the role of the Party was merely to fulfil its leadership function and refrain from meddling in everyday politics. The executive in this scenario would also concentrate solely on its administrative tasks.[387] As the RTFR progressed, documents began to stress the necessity to 'transform government functions' (*zhuanxing zhengfu zhineng*),[388] separating them not only from the Party but also from business.

An official commentary actually suggested taking the reform much further. It opined to disenfranchise townships and towns of their collective enterprises (the aforementioned TVEs) and to outsource administrative functions that could be better performed by private service suppliers.[389] This would result in the reduction of the reliance on extra-budgetary revenues which, as has been noted, derived mainly from fees and levies. Remarkably, it championed to completely abandon the extra-budgetary and extra-systemic revenues and to have all revenues and expenses run through the regular budget.[390] Combined with a fiscal transfer system (*zhuanyi zhifu zhidu*), the gap between administrative responsibilities and available fiscal resources would thus be closed. However, the actual policy did not tackle these issues. It did mention that it was desirable to cease the practice of setting achievement targets, which was, as Chapter 2 showed, an important standard for evaluating cadres as well as a principal source of township indebtedness.[391] However, this and future documents did not follow up on this issue. An institutionalized financial transfer system, as Chapter 6 will show, was set up only two years into the reform.

However, the clearing up of existing township- and village-level liabilities was taken more seriously, and a first step taken to lighten the expenditure burdens of township-level governments was to grant them 'in principle' the receipts of the agricultural tax. It was also decreed that the fiscal relations between the sub-provincial governments should be clarified. However, this issue was not linked to China's national fiscal system, so the call to fund administrative mandates was something of a formality. By stating that poor counties can be subsidized by the province or city, the centre shed its responsibility of being part of the envisioned transfer system. The two points that became the core issues of the Complementary Reform, however, were the substantive reduction of personnel (*jingjian renwu*) and of administrative departments (*jingjian jiegou*) at the county level and below.[392] Among other measures, the document promoted the simultaneous holding of related government and Party responsibilities by one person. In addition, the merging of rural schools and the reduction of teachers were specifically mentioned. Finally, the need to establish a 'supervisory system for burden reduction' was stressed as well. Charges for legitimate services were to be made transparent to the villagers, and the supervision by peasants and the 'public opinion' (*yulun*) was to be strengthened. Again, however, there was no indication of how such a system was to be established and what it should look like. Finally, the document called for drafting and enforcing regulations to sanction cadres who illegally increased the peasant burden.

As is frequently the case in such documents, the final section of the document mandates the 'strengthening of leadership' over the reforms. Rural Tax and Fee Reform Leading Small Groups and corresponding Bureaux were to be established at all administrative levels, and the highest-ranking cadre (*yi ba shou*) should be involved. It is made clear that the provinces were to realize the aforementioned aims by formulating concrete policies after 'broad consultation with grassroots [governments] and peasants'. Finally, the document called on cadres to propagate the policy among the peasants to enable the latter to 'utilize the law to protect their legal rights and interests'.

Impact on central–local relations

As the analysis of the document has so far shown, altering how the refinement and implementation of contentious rural policies are controlled by the central level was not part of the agenda. On the contrary, competition under hierarchy, which was found to bear great responsibility for the peasant burden problem, was also the steering principle underlying the reform that was supposed to improve the peasants' lot. A change of the structural dimension of central–local relations, however, was clearly envisioned.

Regarded through the lens of the three structural dimensions of central–local relations outlined in the previous chapter, it is not difficult to see that these measures, if successful, would significantly tip the balance of central–local relations towards the centre. As for the administrative division of labour, the functions of the county-, and more so the township- and village-level, administrations were being more clearly delineated by the centre, and their internal structure made subject to central regulations. However, regional and local governments were given an explicit mandate to design specific policies in order to realize the abstract aims formulated by the centre, so administrative delegation continued at least with regard to the RTFR. In addition, the strict ceiling on the number of personnel on local government payrolls was declared to be enforced. Furthermore, detailed and strict criteria for cadre hiring, evaluation, remuneration and promotion were part of the reform as well.[393] In the fiscal realm, the scrapping of the various taxes, levies and fees would drastically reduce county, township and village fiscal autonomy. Moreover, a new system of fiscal management and supervision was envisioned. The little budget left at the village level (now mainly the money raised by the villagers themselves) would be managed by the township (*cuncai xiang guan cunyong*). In turn, township-level finances are made subject to management by the county (*xiangcai xianguan*).[394] A more recent policy closes the chain: in the future, the county budget shall be supervised directly by the provinces, bypassing the prefectures (*sheng guan xian*). This has been probed since 2002 in Zhejiang, Guangdong, Henan, Liaoning and Hubei provinces.[395] Finally, the formalization of central–local relations by the RTFR would put a check on local-level discretion in policy-making and would therefore reduce the degree of informal political decentralization. This applies to rural fiscal reforms as well: the passage of a nationally unified RTFR not only eliminated existing local-level experiments, but also subjected further experiments to State Council approval. In sum, the reform envisioned a substantively more centralized and tightly regulated system that strengthened the provinces and counties at the cost of the cities, townships and villages, and therefore increased the centre's leverage over policy implementation.

Also important for the subsequent analysis is the observation that the relationship between burden reduction (Main Reform) and recentralization (Complementary Reform) was not intended to be as complementary as one might at first believe. A closer look shows that the implementation of the centralization measures was in fact a precondition for burden reduction:

> After the Rural Tax and Fee Reform, reduced revenue will affect expenditures. The county and township-level governments must solve this problem mainly by transforming government functions, simplifying the administrative structure, reducing the number of persons supported by public finances, and readjusting the expenditure structure.[396]

Sustained burden reduction thus hinged on the localities' proactive contribution to their own disarmament. If the localities did not implement the policies, the peasant burden would not be reduced. In this way, the centre played peasants and cadres out against each other in a zero-sum game in which the peasant's gain is the local government's loss, and vice versa. Given the fact that the existence of the peasant burden was portrayed as the direct consequence of local government greed, the cadres were on no moral grounds to resist the implementation of the policy. This and the centre's continued manipulation of the tensions between peasants and cadres put enormous pressure on the local governments.

Conclusion

This chapter has yielded several striking and unexpected results. First of all, agency mattered. The policy formulation process must be considered rather brief given the saliency of the topic, and only a very limited number of actors were involved. The last chapter has shown that local governments were obviously more interested in rural fiscal reforms than the central government. While different policies fermented at the local level for almost five years, central leaders remained rather uncommitted. After the death of Deng Xiaoping and the accession of the new leadership, things changed. From the outset, rural fiscal reforms were an item on the new Premier's agenda, though also not very high in priority. This is confirmed by the implementation of a new Grain Purchase Regulation which spelled the end of nearly all local experiments. However, after Jiang Zemin had publicly expressed his support for the Fuyang reform experiment, the issue soon became an issue of highest priority, and the further-reaching Rural Tax and Fee Reform was favoured over the more limited Tax-for-fee reform which Zhu Rongji had previously championed. The choice of Zhu Rongji as Premier had not been uncontested, and Zhu probably needed to show some achievements quickly especially after having been humiliated by the NPC's first ever rejection of a government-submitted legal draft in 1998 (see above). Therefore, his pushing for the early implementation of the reform had political reasons. This confirms the point made in the previous chapter: the timing of the reform was not the result of instability in the countryside.

Second and relatedly, the analysis illustrated how quickly the rural reforms moved from networked bottom-up innovation to top-down competition under hierarchy. Given the speed with which Document No. 7 was drawn up, the similarity of the Main Reform to the policy proposals formulated by Anhui's provincial government, and the small size of the RTFR LSG, there is hardly any doubt that policy formulation was hasted. In order to speed up the process, the Anhui model was simply taken as a blueprint for nationwide implementation, without first deliberating its applicability to other provinces. Hence, the scope of actors changed and national-level policy formulation hinged no more on local actors, but on the Premier and the Group of Three. Adaptability and sustainability were thus sacrificed for personal considerations.

Although agency was mainly responsible for the unfortunate timing of the reform, reform implementation again followed the principles of competition under hierarchy, which leads me to the conclusion that this phase of the reform process was after all institution-heavy. Continuity is apparent regarding the narratives with which the reform was justified, and in the measures eventually taken. In particular, the interpretation of the peasant burden problem followed the narrative of the peasant burden agenda outlined in the previous chapter. It blamed bad governance squarely on local leadership deficiencies and made only passing mention of the more fundamental institutional problems discussed in Chapters 2 and 3. As a consequence, the reform was mainly targeted at reducing local autonomy. Competition under hierarchy, however, remained intact, and the fundamental mismatch between administrative responsibilities and the lack of formal fiscal autonomy at the county and township level was even increased. In fact, the RTFR policies themselves displayed exactly the same competitive character that had so severely heightened the operating costs of central–local politics. The centre again provided reduction targets to achieve, but left it to the provincial and local levels to compete in working out the details. Such preconditions do not bode well for policy implementation, and as the next chapter will show, the reform indeed encountered numerous difficulties and had to be abandoned eventually.

6 Reform implementation and ad-hoc policy innovation

In no phase is the effect that the interplay of contexts, institutions, incentives and agency has on the outcome of a policy more uncertain than during its implementation. Agenda-setting and policy formulation are very controlled processes involving only a manageable number of actors, but when the policy is implemented, a much larger cast of actors is involved. Furthermore, while the policy-formulators are intimately connected to the narratives they are creating and the rules they are passing, the implementers are faced with regulations they had no part in creating. Here, the difference between the solution that would have been necessary to solve the problem (if such a solution could be specified) and the policy that is the product of contending narratives and formalistic compromises becomes most apparent. In the RTFR this is even more the case as there was next to no prior consultation with the implementers, and implementation was hastened so much that careful planning was foreclosed. This chapter traces how the combination of a flawed policy and competition under hierarchy led to the outcomes sketched in the introduction to this volume. Although the policy itself bore scant incentives for enthusiastic implementation, cooperation with the central government did. Opportunities existed to 'fix' the policy in the course of implementation and thereby make it more attractive.

The chapter starts by illustrating how the Anhui provincial government sought to adapt the reform to local circumstances and how implementation ran into severe difficulties right at the beginning. Remarkably, the central government ordered the expansion of the reform nevertheless, hoping that local pioneers would contribute to 'fixing' the policy and thereby lessen resistance. The first implementation phase ended when the central government yielded to local pressure and increased the transfer payments to Anhui Province. The second implementation phase entailed a further fine-tuning of the policy and its extension to all over China. In the third implementation phase, frequent re-regulation took place but did not lead to the desired outcomes. Although the reform did claim some success in peasant burden reduction and administrative reforms, its outcomes lagged far behind the initial expectations. In order to highlight the crucial role of pioneers and resisters in these two stages, their respective roles will be illustrated by drawing on several case studies. The abolition of the agricultural tax between 2004 and 2006 comprises the third implementation phase that will be examined in the last section of this chapter.

Anhui Province pilot phase

Since the RTFR programme had originated and was pre-tested in Anhui, the province was a natural candidate for its pilot implementation. According to Zhu Weixin, bureau chief of Anhui's RTFR Leading Small Group, 'very many policies contained in Document No. 7 are based on Anhui's experience', and granting it trial status was a mere formality.[397] Indeed, a fortnight before 'Document No. 7' was passed in early March 2000, the provincial government had already set up a RTFR Leading Group. On February 28, the Standing Committee of the Provincial Party Committee passed the 'Implementation Plan for the Rural Tax and Fee Reform Trial in Anhui Province'. And two weeks before Anhui was formally granted trial status by the State Council, Vice-premier Wen Jiabao had visited the province for three days. He 'inspected' (*diaojiu*) the reform progress in Huaiyuan County and attended the 'Meeting for the Exchange of Experiences in Economic Work and the Mobilization for the Rural Tax and Fee Reform Pilot'.[398] There, he held a speech in which he announced that 'the centre has designated Anhui as the only province in the country where the reform is implemented province-wide. The Central Party and the State Council place great hopes in Anhui'.[399]

Fine-tuning the reform

Accordingly, Anhui was well prepared for the task. Within three days after Anhui's designation as pilot province, the city and county-level governments were informed that the RTFR had fully commenced.[400] Less than one month later, on May 16, the provincial government handed down nine Supplementary Documents to guide reform implementation at the local level. The following items were regulated: (1) financial system reform of township-level governments, (2) and (3) agricultural and special products tax collection, (4) integration of expenditure items formerly financed by the township 'comprehensive fee' into the township budget, (5) village-level money and labour pooling for village infrastructure projects, (6) administration of village-level finances, (7) standardization of village administration and village-level institution building, (8) tax administration, and (9) procedures for dealing with policy violators.[401]

However, despite their considerable length, all these documents were rather vague in content. Basically, they mirrored the central government's strategy of passing the responsibilities on to the next lower level by terming it 'implementation in line with local circumstances'. Most important, despite the foreseeable loss of revenues to the counties and townships, no regulations concerning subsidies were provided; and neither were there concrete regulations for staff reductions. Hence, the centre's commands were simply handed down, bringing the pressure to bear on the lowest levels. For example, county-township fiscal relations, one of the key structural problems in China's fiscal system, were to be dealt with in the following way:

> Consider all factors while planning, appropriately address the imbalance in township and town finances [*caizheng qingxie*] Adjust correctly the fiscal relations between county and township; sensibly reallocate county and township fiscal resources. While ascertaining that the fiscal needs of the counties are met, take utmost care to protect the fiscal interests of the townships and towns, and ensure that the basic expenditures needed to keep township and town organization functioning are met.[402]

In other words, these documents did not tell the subordinate levels what exactly they are supposed to do, but to do it properly and correctly.

However, the basic parameters of the reform were in place. As the administrations of all levels undoubtedly knew, the abolition of all non-service fees, of the slaughter tax and the township comprehensive levy spelled extraordinary fiscal losses for most of Anhui's counties and townships. Yet neither the centre nor the Province had passed regulations concerning subsidies, although Document No. 7 explicitly demanded them. Thus, there was a gap between the dire need to drastically reduce county- and township-level expenditures (or increase revenue) and the lack of regulations of how to achieve this goal.

There are indications that counties sought to avoid this situation by modifying these parameters at a very early stage of reform implementation. The uniform application of the new tax rules would, for example, have reduced F County's revenue by almost one fourth. Therefore, when F County submitted its new tax quotas to the Provincial Ministry of Finance, it pleaded to reduce exactions only a token 6 per cent. It argued that there was no need for more drastic reductions because the county 'had strictly followed the burden reduction rules set by the centre, the State Council, and the province', so 'the overall level of the peasant burden has always been low'. Also, it expressed worries that the 'regular exercise of political power' could not be guaranteed anymore if the reform was implemented to the letter. Specifically, it posited that 'revenue of those counties where the village retention formed a large part of the *santi wutong* will be diminished greatly'.[403] There is some truth to this reasoning. No matter how big the share of the village retention in the *santi wutong* was before the reform, it could not exceed 20 per cent of the new agricultural tax thereafter. Thus, the leadership of F County reasoned, counties with high overheads for village administration and social services would face disproportionately high difficulties. In addition, the County complained that fiscal pressure had been huge even before the reform. Finally, it had just repaid '82 million yuan in fiscal and trust funds owed to the province and the city' in 1999.[404] Hence, the reform (or the demand to repay the funds) came at a very bad time. F County complained that it had already been short-changed in the tax reforms of 1985 and 1994, and was now being punished for having observed central and provincial burden regulations.[405] Had they taken more money in the past, the reform would now be easier to swallow. The pleas, however, went unheeded.

F County later reported to have reduced the peasant burden more than 20 per cent.[406] In fact, 20 per cent can be considered the minimum margin of burden reduction. Only three of sixty-one counties had fallen below this threshold and one of them was Huaiyuan County, which was one of the pilot Counties that had already implemented the reform one year earlier. Asked how the local level dealt with the problem of revenue shortages, the former head of the RTFR small group in Anhui's W County neatly summed up the corridor of incentives: 'This policy had highest priority and absolutely had to be implemented. As for the lack of revenue, you had to find a way how to solve this problem. However, you were under no circumstances allowed to increase the peasant burden'.[407]

The provincial government took great care to inform the peasants about this fact. Most notably, the Office of the provincial RTFR LSG delivered a 'Letter Addressing the Peasant Masses of the Whole Province' to each peasant household. This letter explained both the policy and the rights and responsibilities of the peasants very clearly, and gave them hope that their complaints would be heard:

> 'With the implementation of the Rural Tax and Fee Reform, peasants engaging in agriculture will not be paying any administrative or service charges or fund contributions. The only exceptions are the agricultural tax and its surcharge; the tax on agricultural special products and its surcharge; regular fees set by national laws or regulations and formally promulgated and retained after consideration by the provincial government; and fees related to production, such as fees for water and electricity. As for the taxes and fees that the peasants are required to submit, the departments in charge will inform the peasant masses by means of a Tax Notification and a Peasant Burden Supervisory Card. Peasants have the right to refuse payment of any administrative charges, fund contributions, etc., which are not covered by the Tax Notification and the Supervisory Card, and rely on the law to protect their legitimate rights and interests.'[408]

In addition, the reform was widely publicized in the newspapers, advertised on Provincial TV channels, and spread by word of mouth. Hotlines for lodging complaints were installed. In addition, the reverse side of the 'Tax Notification Card' carried the phone numbers of seventeen Provincial departments responsible for the administration of taxes and fees.

After the Summer and Autumn harvests, when the new tax had first been collected, the higher levels started pushing the implementation of the Complementary Reforms. On December 26, the General Offices of the CCP and the State Council decreed that administrative personnel in the city, county and township bureaucracies were to be reduced by 20 per cent within three years; the concrete arrangements were once again to be decided by the provinces 'according to local circumstances.'[409] Three months earlier, however, the Anhui provincial government had already passed its own rules concerning personnel reduction, demanding the reduction of only 10 per cent of township/town-level administrative personnel and making no mention of reforms at the county and city levels.[410] Despite the passage of the central document, this was not corrected. The same central document demanded that the localities continue paying the salaries of demoted personnel for another three years. Temporary employees would be paid a severance if their employment had been authorized by the centre; all the others would receive nothing. The document stated that the centre would jump in for these expenses if localities were in financial straits (which almost all of them were), but no transfer payments were earmarked specifically for personnel costs.[411] The transfer payments the province did receive were too little (see below) and in many cases did not reach their destination.[412]

Unintended consequences

According to Zhu Weixin, the reform's 'initial costs' came to 5.4 billion yuan for the whole province.[413] The loss of income due to burden reduction alone totalled more than 2 billion yuan, and subsidies to townships and villages another 1.2 billion yuan.[414] In addition, a substantive part of local government revenue had come from irregular exactions, which now were illegal. For thirty-five of sixty-one counties, the agricultural and special products taxes, the village retention, the township comprehensive levy and education fees had equalled more than 50 per cent of their budgetary revenue.[415] Although the reform caused severe shortages for more than half of all counties, the central government was slow to react, probably hoping that Anhui would be able to cope with the situation by itself.

This shortage of funds very quickly had disastrous impacts on rural education. As noted in Chapter 3, rural compulsory education was primarily financed by the townships and villages, and overwhelmingly with non-budgetary resources. Since 1994, province-wide arrears on teachers' wages had added up to 1.7 billion yuan,[416] and the RTFR was bound to worsen this situation. In 2000, Anhui's townships and towns received only 4.6 billion yuan in funds, but needed 4.95 billion yuan alone to pay the wages of 660,000 employees. Hence, wage arrears became even more serious. In Fuyang City, for example, teachers had on average not been paid for ten months.[417] In general, township- and county-level cadres were reluctant to fire employees, and many complained that they were not given enough time.

These problems were aggravated by the fact that many peasants refused to pay the new agricultural tax, which further diminished local budgets. The reason for this boycott was that the tax proved to be both highly unfair and in some cases the object of manipulation. This was again rooted in the mode of tax administration: the township had to report the amount of *registered* farmland in its jurisdiction to the county, along with the average yields between 1993 and 1998. The county then calculated the sum that each township had to deliver, based on market prices. This sum was then divided by the *actual* acreage farmed in all villages. Each farmer had to pay a fixed amount of money or grain for every *mu* of his contracted land, regardless of the yield and grain planted. However, due to several factors, for example the requisitioning of land for public projects or simply over-reporting, the amount of registered farmland was nearly always higher than the actual acreage. Accordingly, the farmers had to pay more than they were supposed to. Also, not being progressive, the problem of taxes consuming a far larger part of the net income of poor peasants persisted. Finally, some refused, were not able to pay or were exempt from paying the tax, and some were simply absent, as in the case of some migrant workers. Hence, it was frequently impossible to raise the full sum that had to be delivered to the county. Therefore, the missing amount had again to be apportioned among the tax payers, which further raised their obligations towards the government. Often, the village had to use its share of the new tax, the 20 per cent supplement, to meet its tax obligations. As a consequence, village-level service provision began to stagnate where other sources of revenue were absent.[418]

For these reasons, fee-taking continued in many localities. During his field trip to Anhui's Suixi County, Chinese researcher Gao Xinjun found that six out of twenty-four towns were forced to continue collecting irregular fees already in the first reform year because they could not make ends meet.[419] As a result, the number of letters and complaints filed by peasants increased extraordinarily in the first years of the reform, which Figure 6.1 clearly illustrates.

Decision to expand the reform

These problems had probably not been foreseen by the centre, and if they had been, they did not prompt the leadership to rethink the reform. In hindsight, it seems that some key decision-makers were uninterested in the policy, while others were ignorant of the problems. As the following paragraphs will show, the Ministry of Finance was most concerned with keeping the costs of the reform low. The Premier, on the other hand, wanted the policy to succeed and was flexible with regard to costs. However, he was quite unaware of rural China's problems. The Ministry of Finance's passive role in the process is confirmed by a 'veteran provincial official' cited by Linda Chelan Li:

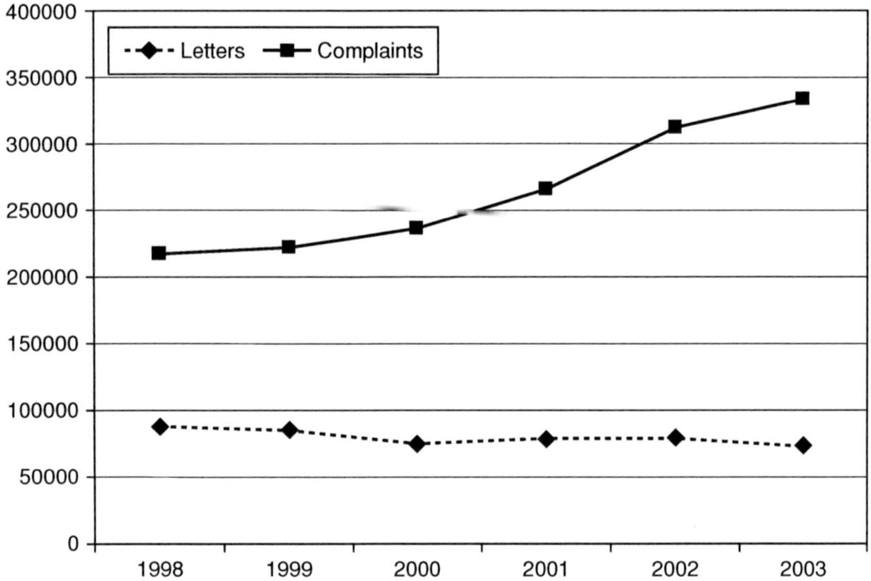

Figure 6.1 Number of letters and complaint visits to county, municipal and the provincial government, Anhui Province, 1998–2003

> The Ministry of Finance, in charge of the reform since 1998, had, for instance, little expertise nor interest in rural governance and the peasantry. They had thus adopted a pedantic approach to the issue of reducing the peasant burden, seeing it narrowly as a question of improving fiscal implementation and rationalization. They had not taken the broader picture of the national fiscal system and rural–urban resource allocation into consideration.[420]

The events surrounding the planned extension of the reform support this assertion. Already in early November 2000, Minister of Finance Xiang Huaicheng held a speech at the Symposium for the Nationwide Work on the Rural Tax and Fee Reform.[421] He signalled that the reform would be extended nationwide, and that the centre would allocate 20 billion yuan annually, and not much more, to support the regions. In fact, he almost defiantly told the provinces to complain less and work harder:

> Comrades from several places pay rather close attention to the amount of transfer payments from the centre. They report difficulties and demand higher increases in monetary transfers. But they talk much less about how to improve the foundation of the reform in their locality and how to proactively and appropriately implement it, or how to apply complementary measures to solve the difficulties and problems which

> have surfaced. It is understandable that everyone wants the centre to increase monetary transfers, and it is correct to seek the truth from the facts and report difficulties and make demands. However, the centre's transfer payments are merely one of the important preconditions that push forward this reform. The key factors are if the centre's guiding thoughts, aims and measures regarding the Rural Tax and Fee Reform are understood and supported by cadres of all levels and the peasant masses alike; if the Provincial Party Committees and goverments take the firm leadership in the reform work; if the programme meets the expectations of the majority of the peasant masses; if the peasant burden is reduced, and if the conditions are met to ensure the continued functioning of the grassroots governments. In one sentence: the key lies in the quality of our work. If the implementation programme is not thought through circumspectly and precisely, if there is no preparation, if the majority of the cadres and peasants cannot be mobilized to actively embrace the reform, if we do not actively and avidly plunge ourselves into the reform work, then it does not matter how much the central government will transfer, because problems will always surface.

Given that Complementary Reforms had not been implemented yet by the localities, Xiang implied, the centre had been very generous:

> The State Council has decided that 20 billion yuan from the centre's coffers will be earmarked annually as transfer payments to support implementation of the Rural Tax Reform across the country. This fully shows the commitment of the Party Centre and the State Council to push forward the reform and reduce the peasant burden. It is also a concrete manifestation of its support to agriculture, of protecting the peasants' interests, and of being concerned about political work at the grass roots. All provinces are extraordinary sympathetic towards the centre's fiscal situation. They are of the opinion that it is not easy to support this reform by providing so much funding when central finances are so deeply in the red and debts are so huge.

He made the message of his speech very clear: reform pilots could draw up cost estimates to account for factors such as a reduction in arable land or fishable waters because of construction projects or environmental factors. Of course, Xiang stated,

> there are some localities who, when applying for fiscal help from the centre, base their calculation not only on the township comprehensive levy and the village retention, which are legal exactions according to national regulations. Rather, they also figure in several of the wanton levies that they had illegally charged before the reform. For several years, the Party's Central Committee and the State Council have taken very

> seriously the problem of the 'three wantons', and they have passed down a series of documents and adopted a series of measures, they have strived to bring the problem under control, and their aim just was to end all kinds of wanton fee-taking and to prevent the peasant burden from increasing. If we recognize these wanton items that were drawn up against central regulations, if we add them into calculations of the fiscal deficit of the locality, then we would allow these rule-breakers to profit, while those that have followed central regulations would get less and suffer losses. This does not conform to the spirit of the centre. Hence, we cannot take into consideration those deficits that have been incurred because of wanton fee-taking.

In mid-February 2001, the *Xinhua* news agency proclaimed that reform pilots would be spread across the whole country that same year, and that the RTFR would become a national policy by 2002.[422] Four days later, a conference was held in Hefei, the capital of Anhui Province, to discuss these aims. The conference was attended by the leadership of the twenty provinces that had implemented the reform in selected localities, the relevant Ministers of the State Council, and forty-eight Ministry-level provincial cadres.[423] Xiang Huaicheng basically reiterated his points made three months earlier, though less fervently: The centre would provide 20 billion yuan in transfer payments, and the provinces were exhorted to increase their subsidies as well. In addition, they were to 'coordinate their subsidies with those of the centre and implement a system for administrating transfer funds in order to guarantee that the transfer payments are effectively used. Localities in fiscal difficulties must be appropriately considered'. However, 'if post-reform rural fiscal deficits are met merely by means of fiscal transfers, the reform has lost its meaning'.[424]

Premier Zhu Rongji saw the matter somewhat differently. When asked at a press conference on March 15 about the difficulties of the reform, he recognized that the central government could not simply ignore the illegal levies just because they were illegal. He stated that 'twenty to thirty billion yuan will be allocated from central coffers to subsidize the villages and poor provinces, autonomous regions and cities directly under the jurisdiction of the central government. However, the deficit is still very large'.[425] When a journalist later pointed out the difficulty of reforming rural education under such circumstances, Zhu remarked: 'We absolutely want to fulfil the aim of guaranteeing rural compulsory education. If there is not enough money, we will just add money.' However, he was reluctant to provide concrete figures and told the journalist to ask him again the following year.[426] As planned, a circular was passed a fortnight later to set up pilots all across the country. Less than one month after that, however, the General Office of the State Council suddenly ordered the temporary cessation of the expansion (though not of the work in the existing pilots).[427] The central government had either run out of money,[428] or, what is more likely, it had to decide if the reforms were worth the cost. According to Linda Chelan Li, the provinces had reported in April 2001 that they needed 100 billion yuan annually from the central government.[429] Most probably, the central government had to decide how deeply it wanted to get involved. It took about one year for the reform to restart, a delay that Linda Chelan Li attributes to 'central-provincial bargaining over reform costs'. She concludes that

> elite competition and bureaucratic boundaries worked against proper reform preparation at the central level, despite a long gestation at the local levels and availability of better knowledge elsewhere. The resulting reform package reflected a very partial, and lopsided, understanding of the issues involved, and aimed at making local governments absorb most of the costs of reform. Provincial governments were naturally opposed to this and therefore lobbied hard for central transfers.[430]

Crisis and the hard hand

While this undoubtedly is a very accurate observation, it was not only central–local bargaining that caused the delay. Very probably, the high projections reported by the provincial governments had alerted central-level leaders that the reform experiment was not going as well as Anhui had reported. The high number of reform-related provincial-level documents passed between April 2001 and March 2002, when the reform was continued, suggests that the Anhui Provincial government was told to improve existing flaws before the reform could be implemented nationwide. Anhui had passed only ten such documents between the initiation of the reform and April 2001. Eight of them clarified the basic parameters of the reform and basically restated the content of eight central-level circulars passed during the same period. Standards were set for agricultural tax collection and administration,[431] for abolishing illegal and making transparent legal administrative and service fees,[432] for improving the 'Burden Supervisory Cards' system,[433] for trimming township and town governments,[434] for rural development,[435] and for assessing the agricultural special products tax.[436] The other two were 'Urgent Notifications' passed in late 2000, which already implied that illegal tax and fee collection was getting out of hand and that there were severe problems with rural service provision.[437] Between April 2001 and March 2002, only one central-[438] but twenty-one provincial-level documents were passed, many of them substantial and quite, but not too, specific. These documents addressed four groups of deficiencies and are therefore a good indicator of the implementation lags in the first reform stage. Specifically, they indicate that taxes were assessed on an unrealistic base,[439] that fee-taking continued especially, but not only, in the education sector,[440] problems with the provision of compulsory education and other social services,[441] and that the liabilities of grassroots-level governments had further increased.[442]

A circular passed on May 16, only shortly after the reform experiment was frozen, is very instructive with respect to the deficiencies and how they were addressed.[443] It contained five supplementary documents aimed at improving tax assessment and–collection, financial support for the poorest, and financing grassroots-level services. As for taxation, two of the supplements reiterated instructions given earlier, indicating that they had not been followed. The township-level leadership was now explicitly made responsible for the correct assessment of local tax duties, and the supplements forbade the manipulation of the two tax assessment parameters, which were the amount of contracted agricultural land and the average harvest between 1993 and 1998. Idle and unusable land was not to be figured in, and the peasants were encouraged to contest incorrect figures. Tax collection was to be 'regularized' (*guifanhua*). An open, transparent and contestable tax register stating each household's duties was to be set up. In addition, each household should receive a written tax notice, and taxes were to be collected at a designated locality and at a fixed time. Tax officers were to be selected by means of examinations and to receive special training. Finally, the document decreed that 'each locality should set up a sound (*jianquan*) tax collection and administration organization to guarantee that state taxes are regularly collected, and that the agricultural tax enters the state treasury in sufficient amount and timely fashion.'

The rights of households entitled to 'special protection' (*teshu baozhang*) were also reconfirmed. Elderly and handicapped people who were unable to work, had no source of income, and who had no legal guardians whatsoever to support them continued to receive food, clothing, housing, medical care and burial services (the five guarantees; *wu bao*) from the collective. Part of the township comprehensive levy had previously been used to finance these expenditures.[444] Also entitled to government subsidies were the dependents of conscripted soldiers (*yiwubing*) and households afflicted by 'special difficulties' (*teshu kunnan hu*).[445]

This takes us to the financing of grassroots-level social and other services. In principle, funding now came from two sources. Public projects from which all villagers profited could be financed by the 'one undertaking, one decision-making' method explained in the previous chapter, but all other expenses had to be borne by the township-level governments. The same is true for the selection, training and evaluation of tax officers, and the collection and administration of the agricultural tax. Although the fiscal departments of all levels from the province down were responsible for administering the tax, the setting up of a 'sound tax collection and administration organization' solely became the townships' duty. In short, the supplements unrealistically stipulated that social services should be improved and tax collection regularized, but without increasing the peasant burden.

In early September 2001, another policy package was handed down which addressed different issues but followed a similar logic.[446] With regard to service provision, one of the supplementary documents voiced concern about 'rural middle and primary schools in danger of collapse'. Specifically, schools in danger of collapse were to be renovated or torn down 'within about two years' from 2001, and the construction of new 'dangerous buildings' was to be prevented. However, another supplement decreed that villages and townships were not allowed to incur new liabilities without offering sufficient collateral, and that they could not anymore serve as guarantors for enterprise loans. In addition, they were forbidden to borrow from individuals and non-bank organizations, and to borrow at high interest. Finally, existing liabilities could no longer be apportioned among the farmers. One supplement specifically extended this rule to World Bank credits, whose repayment obviously had to be shouldered by China's farmers before the reform. A final document addressed the fiscal relationship between townships and villages, forbidding the former to retain the surcharge to the agricultural tax and subsidies intended for the villages. It complained that 'some places use the surcharges to the two agricultural taxes and village-level fiscal subsidies to meet their tax quotas or repay their liabilities . . . or even misappropriate them' and thereby endanger the 'regular operation of the village administrations' and, in consequence, rural stability.

During the same period of time, the authorities stepped up their efforts to enforce the implementation of these documents. With both the foreign and the national media predicting the failure of the 'third rural revolution', the pressure on Anhui Province (and the centre!) was tremendous. In late September 2000, an 'emergency video and telephone conference' was convened by the Anhui provincial Party Committee and the government. The purpose of this conference was to '. . . address the serious contradictions that have emerged during the Rural Tax and Fee Reform and the extraordinary rise of collective petitions that have accompanied them.' The outcome of this conference was a 'responsibility system' for the work in letter and complaints. Those leading Party and government cadres who 'did not take seriously' such work would be 'punished'. Depending on the severity of the situation, the result would be 'criticism in an internal circular', 'an instruction to write a self-criticism', 'a warning of dismissal from office' or 'punishment by the Party or government disciplinary committee'.[447] The latter usually resulted in the dismissal from office or, in the worst case, the revocation of Party membership and thus the end of one's political career. These were no empty threats. According to Zhu Weixin, bureau chief of Anhui's RTFR Leading Small Group, the government had been decidedly heavy-handed in their efforts to ensure the policy's success:

> We dealt with them [cadres violating the RTFR policies] severely. Single cases involved more than ten, twenty, several dozen people. If we have to punish (*chuli*) them, we will punish them. The main aim is to ensure the continuity of the policy. . . . We were able to improve some cadres' conduct by ways of educating them. We do all we can to not have to punish people.[448]

Most cases occurred in 2000 and 2001. By 2003, 822 cases had been investigated and 224 people punished, most of them township-level cadres. By 2004, according to Zhu, both the number of such cases and the number of people who petitioned the provincial government due to irregular fees had declined drastically. 'That year, there basically were no cases'. This might very well be possible, but should not be equated with the success of the reform, at least not as it was originally intended. First of all, the Anhui provincial government was forceful in disciplining the cadres. In itself, this did not solve the problem, but merely contained it until government vigilance subsided. Second, the actual number of cases was much higher. As the former head of W County's Policy Research Commission pointed out when confronted with these figures, 'people were punished pretty much everywhere, but these cases were never recorded.'[449] Third, the centre had declared in March that year that the agricultural tax would be scrapped within six years, and the special products tax even sooner. The peasants probably could have chosen to wait these changes out. Finally, transfer payments from the centre had started to pick up (see next section).

As these developments make clear, the year 2001 marked a watershed in the reform. The reform had not changed yet the institutional set-up that hinged a cadre's political fortune on fulfilling fiscal responsibilities vis-à-vis the upper level. At the same time, however, local governments were expected to abandon illegal levying. Given the fact that cost-saving measures had not yet been widely implemented, these incentives were mutually exclusive for many cadres. Often, they chose to risk disciplinary measures over being accused of underperformance, because the former not necessarily meant demotion, whereas the latter did. Township- and village-level cadres were hit hardest by this dilemma: they were at the bottom end of the administrative hierarchy and could not delegate their responsibilities further down.

Transfer payments

As the reform was unfolding, the centre continued to hand down documents that more or less directly addressed the problems that had sprung up. Although Document No. 7 had ordered the localities to finance the overwhelming part of the reform by implementing the Complementary Reforms, the Ministry of Finance five months later decreed that a 'standardized subvention system' would be 'gradually instituted.'[450] However, it took the Ministry two years to even lay out the foundations for such a system.[451] Burden reduction had been going on for two-and-a-half years before standardized measures were in place to deal with the losses to local government, a lag which wrought havoc on county finance and made the counties highly dependent on financial infusions from above and remittances from below.

Nevertheless, irregular and small-scale transfer payments had started as early as 2000. That year, the central government transferred 1.1 billion yuan to Anhui. The provincial government added another 100 million yuan, so altogether 1.2 billion yuan were available to offset the losses caused by the reforms.[452] However, this only covered about two thirds of the deficit, and far less if the significant, yet unrecorded irregular exactions are figured in. This loss predominantly affected the township level, where 83 per cent of the deficit occurred.[453]

Why did the counties and townships initially not get rid of surplus employees? One reason cited by three interviewees was that the leading cadres were reluctant to fire employees, as many of them held their position as a result of patronage politics.[454] Another reason was that local-level cadres refused to shoulder the reform costs alone and expected the centre to provide generous subsidies. The head of W County's RTFR Small Group put it bluntly: 'It's the centre's reform, so the centre must pay for it'.[455] So how can revenue be increased without exploiting the peasants, without getting rid of employees, and in the absence of rural industry? The same cadre pointed out that the success of the RTFR had highest political priority, and he implied that an inverse relationship existed between a policy's priority and the firmness of budget constraints. As had been the case with other target-setting programmes, township-level governments threw themselves further into debt in the initial phase of the reform.[456]

As the above paragraphs suggest, there seems to have been considerable disagreement on who was to shoulder the substantial costs of the reform. In what amounted to a tug-of-war, both sides were waiting for the pressure to build up. The centre had hoped that revenue shortages would force the localities to implement the Complementary Reforms quickly, while the localities hoped for increased transfer payments. However, the veteran reform protagonist of W County dismissed the Complementary Reforms as 'unreasonable' and not being rooted in practice and claimed to speak for the majority of local governments.[457] The fact that the Complementary Reforms were not and could not be uniformly implemented, but were also the subject of local experiments supports this view. This and the fact that the centre had grudgingly started to reimburse the Province tilted the balance in favour of the localities. A well reported visit by then Premier Zhu Rongji to Anhui Province seems to have proven the localities right and at the same time to have ended the struggle.

A verbatim record of the Premier's visit to an Anhui primary school illustrates the lack of knowledge of key central-level politicians about rural affairs. First of all, Zhu Rongji was astonished to learn that in an 'average' Fuyang primary school, pupils had to bring their own stools. Upon learning that the school lacked the resources to change out-dated furniture, he reportedly reacted the following way: '[a]fter a good while of silence, Zhu Rongji felt a mottled desk top and said with feeling, 'Very difficult indeed!' Furthermore, he seemed to be thoroughly unfamiliar with the pressures that led rural cadres into levying from the peasants. For example, his lecture about the negative impact of excessive fees and about the necessity to follow central regulations was not well taken. Zhu seemed surprised when he was told that many townships had to take recourse to such fees because they were unable to even pay their teachers. He replied: 'If a township indeed has difficulties in funding teachers' wages due to insufficient financial resources, the problem should be solved by readjusting expenditure structure and by increasing central fiscal transfer payment to local authorities'. He promised to personally make sure that wage arrears would be paid.[458] He held true to his promise: On September 4, 2001, Anhui's leadership decreed that responsibility for the salaries of primary- and middle-school teachers was transferred to the county level.[459]

According to Zhu Weixin, this visit ended the struggle. From then on, transfer payments from the centre gradually picked up. In 2001, the province received 1.7 billion yuan in earmarked subsidies for reform implementation, an increase of 55 per cent from the previous year.[460] However, the provincial and city-level governments also had to increase subsidy payments. In total, 2.3 billion yuan were earmarked as fiscal subsidies for Anhui's townships and villages.[461] Hence, local governments did not have to bear the reform costs by themselves, as initially expected, but a large part came to be shouldered by the centre. Except for a few promising localities where they were tested, however, the Complementary Reforms remained largely unimplemented. A central-level policy adviser also subscribes to the interpretation of this phase as a tug of war: 'the centre was willing from the start to bear a significant part of the reform costs, but did not show it immediately. If it had started handing out transfer payments with the onset of the reform, the local governments would have demanded more and more.'[462]

Expansion phase

After this *tour de force* and despite the still present difficulties in Anhui province, the RTFR was significantly expanded. In late March 2002, the General Office of the State Council issued a document ordering the expansion of trial areas to sixteen other provinces, whose governments were to decide if they wanted to implement the reforms full-scale or only in some trial counties. Despite the numerous problems still present in Anhui province, the circular did not depart much from previous documents and basically repeated the existing policies. However, political learning was reflected in two innovations.

Further adjustment

As shown above, burden reduction had led to grassroots-level inter- and intra-governmental competition for fiscal resources, and had often been achieved at a high cost for rural service provision. The circular therefore narrowed this corridor by proclaiming that these issues were of equal importance. The 'iron triangle' of burden reduction, improved education and smooth grassroots administration was melded into a 'guiding principle' of the reform called the 'Three Guarantees' (*san ge quebao*). Apart from the significant and sustained reduction of the peasant burden, the smooth functioning of village administrations and township-level governments, and the availability of sufficient funds for rural compulsory education had to be guaranteed.[463] The second and related novelty was the central government's implicit acknowledgement that these goals could not be fulfilled without increased financial commitment from above. Four months after the expansion, a standardized fiscal transfer system was finally put into place. With the new system, localities received reform-related transfer payments that consisted of three component parts: township/town subsidy, village subsidy, and education surcharge subsidy. Subsidies for the township level were, as before, calculated on the base of the 'official' losses incurred by the reform, so revenue from illegal levies was not figured in. However, two important adjustments were made. Unlike before, 'other comprehensive expenditures' could be added to the losses, which clearly was a concession to the localities. Second, the resulting figure was multiplied by each township's 'subsidy quotient'. The quotient considered each locality's pre-reform dependence on agricultural taxes and fees as well as wage and administrative expenditures. It was calculated by dividing the percentage of these revenues and expenditures in the local balance sheet by the national averages, adding them up, and multiplying them by a 'central government fiscal burden coefficient'. This quotient was set by the centre and depended on the centre's commitment. Apart from setting this factor, the centre had additional leverage over the formula by specifying a 'relevance factor' (*quanzhong*) by which the two percentages had to be multiplied. Thereby, it could, for example, reward those counties that were highly dependent on agricultural taxes and fees, but had low wage and administrative costs.[464]

In 2002, 24.5 billion yuan were distributed among the twenty provinces that had joined the reform, almost five billion more than the amount promised by Xiang Huaicheng for country-wide implementation (see below). In addition, the provinces received larger tax returns from the centre. While provincial expenditures exceeded revenues by 68.3 per cent in 2001, the ratio increased to 79.5 per cent within just one year.[465] Thus, the reform had become attractive for provincial governments. According to an Anhui provincial official,

> It is pretty simple why we started. The central government promised us additional subsidies to kick-start the reform. Why not snap the new monies? Reform or not, these new resources will add to the local coffers. The provincial government cannot lose out one way or another by joining.[466]

Finally, though not an innovation, the centre further encouraged experiments with administrative and other reforms that would soften the impact of the tax and fee reductions. As the next sections will show, the reform did not only encounter resistance, but was seen by several localities as an opportunity to function as pioneers in certain policy fields. The innovations produced by such pioneers had then to be implemented across China, and resistance was not accepted. Hence, the centre combined concessions, an invitation to cooperate in policy formulation and a hard hand in policy implementation to make the policy more acceptable and crush resistance at the same time. An interesting picture emerges: the reform achievements claimed by the central and provincial government were subject to great local variation, and since false reporting was one of the resistance strategies to be outlined below, these achievements are far less secure than the aggregate figures make believe. As the next chapter will show, this is also true for the increase of peasant income. In addition, the new agricultural tax was neither welcomed by the peasants nor by grassroots cadres, and the necessity to fulfil tax quotas led to great frictions between both groups as well as between townships and counties. Not only did the new policy prove unable to solve the problem of wanton exactions, but the meagre and limited achievements came at a substantial cost to rural service provision. For these reasons, abandonment of the policy and increased central-level subsidies and political steering were probably the only viable answer.

Pioneers

As Jae Ho Chung's study on the implementation of the household responsibility system across China illustrates, three necessary conditions must be fulfilled for localities to engage as pioneers for certain policies. The economic circumstances must either allow or make them necessary, political leaders must be willing to engage in such reforms, and they must have the support of key individuals at the higher levels.[467] Chapter 4 has shown that political support was the key variable that determined the success of the implementation of reform trials in Xinxing Town, Taihe County and Zhengding County. After the RTFR had been formally implemented in Anhui Province, several localities devised schemes for putting into practice individual Complementary Reform polices. Hence, while the 'main reform' had been uniformly implemented by 2002, Complementary Reform trials sprang up across the Province. The following sections introduce two pioneers who experimented with polices that were later implemented province- or even county-wide. Wuhe County first implemented large-scale administrative restructuring and personnel reduction measures, while He County experimented with grassroots-level fiscal reforms.

According to official documentations, Wuhe County in central Anhui province had joined the league of RTFR experimenters in 1998 by implementing the 'Tax for fee' programme. As one senior reformer pointed out, however, Wuhe had actually been among the reform pioneers and had implemented a reform in tax and fee collection as early as 1995: 'Taihe had implemented its reform one year earlier than we did, but only because our preparatory work took longer. In fact, our preparations started at the same time, in 1993.' According to the interviewee, Wuhe engaged in the reform

> because levying from the peasants was multifarious, heavy, and chaotic [*duo, zhong, luan*], gravely endangering the relations between peasants and cadres. We wanted to solve this problem by implementing a tax collection reform [*zhengshou fangfa gaige*]. At that time, you could under no circumstances manipulate the average harvest rate, the tax rate, and the amount of the township and village levies. You could collect the agricultural tax, and you could collect the township and village levies, but not anything else. So we decided to collect them together, but administer them separately [*tongshou fenguan*]. Wuhe County initiated this programme, it was a reform that started from the grass roots [*zhi xia er shan*]. We saw reform as a gradual process and did not want to solve all problems at once.

The idea came from a cadre in the County's Policy Reform Commission, and immediately received the support of the County leadership. The policy shared many features with the 'Tax-for-fee reform model' that was later brought forth by the Ministry of Finance. As had been the case in Xinxing Town, it was believed that by regularizing and making transparent tax and fee collection, peasants would be more willing to pay their dues, and collection costs could be reduced. For this purpose, a predecessor of the 'Peasant Burden Supervision Card' informed farmers about the exact amount of their tax dues. Allegedly, the township-level cadres and also the peasants were very content with the outcomes. However, the Bengbu City authorities forbade Wuhe to propagandize its experiences or introduce them to other localities (*bu xuanchuan, bu jieshao*). According to the interviewee, 'the Taihe reform experiment was being conducted at the same time, and it was much bigger than ours. This was the show project of the province, and a parallel project was uncalled for'.[468]

Two years later, in 1997, Wuhe again played the role of a pioneer when it experimented with a reform of the agricultural special products tax. Initially, the interviewee had proposed to abandon the tax completely, since it was 'very difficult to handle' (*caozuo hen kunnan*). However, one cadre of the Provincial Ministry of Finance alerted the local official that 'this is not practicable at the present time. Once it is abolished, the tax on agricultural special products cannot be reintroduced'. More interesting than the reform proposal is the fact that a dense web of informal channels of communication existed between the County and the Province. During all reform experiments, the interviewee claimed, he maintained close contact to Ministry-level officials in the provincial government. For example, he 'very often' met with Zhu Weixin, who later headed the office of Anhui's RTFR LSG. Zhang Ping, who then was executive vice-governor, also was an important member of the reform network, as was the afore-mentioned He Kayin. Obviously, it was rather the Provincial than the city-level officials who served as a seismograph for the permitted depth of a reform. In addition, the fine details were mostly worked out between ambitious officials at the department and sometimes Ministry level.

Having successfully experimented with local tax and fee reforms for such a long time, the reformers in Wuhe County quickly realized that the reform could not be sustained with present employment levels, and they were among the first to engage in profound structural changes and personnel reductions. By merger, the number of villages and townships was reduced 48.6 per cent and 25 per cent, respectively. Township-level administrative units in Party and government were reduced 79.6 per cent, and service units 51.6. As a consequence, administrative personnel were reduced 10.0 per cent, service personnel 40.6 per cent, and village-level cadres 64.8 per cent. By merger, 195 rural primary and middle schools were reduced, and one teacher was now responsible for 23.1 instead of 15.7 middle-school pupils. On the primary school level, the increase was even more drastic. In the course of the reform, the teacher to student ratio increased 55.0 per cent, from 17.1 to 26.5.[469]

Another trend-setter for complementary reform implementation was He County in the eastern part of Anhui, bordering Zhejiang province. He County implemented a reform experiment that changed the fiscal relationship between the county and the township-level governments. With a programme called 'township finances are administered by the county and used by the township' (*xiangcai xian guan xiang yong*), it reacted to the budget constraints brought about by the reform and in effect brought township-level finances under county control. Zhou Weihan, head of He County's fiscal department, claims that the reform had become necessary because

> revenue from national and local taxes had declined after the Rural Tax and Fee Reform was implemented. At the same time, the agricultural taxes and fees, which had been the main sources of township-level revenue, had also declined at great scale after the reform. Second, the fiscal balance had become very fragile. Absolutely necessary fiscal expenditures had quickly expanded, and wage expenditures had increased especially fast. This greatly disturbed the balance between fiscal revenues and expenditures. Within only two years, the share of wages in the county's total expenditures increased from 73 per cent to 81 per cent. Third, township-level debts skyrocketed after the reform. This had a very disturbing effect on the operation of the townships and towns and has endangered village stability. On average, each town is indebted to the tune of 4.2 million yuan, and one township even has 17.8 million yuan in liabilities! The fiscal difficulties of the townships were caused on the one hand by the central government's macroeconomic policies. The county level's fiscal scope is rather small, and its capacity to help

> the townships is very limited. On the other hand, however, the township level's methods of fiscal administration are crude (*xiangzhen licai fangshi de cufang*), the fiscal system is not sound (*jianquan*), and the budgetary constraint mechanisms are not very strong. The township cadres wantonly issue orders outside the budget (*yusuan wai luan pi tiaozi*), wantonly engage in projects that need increased spending (*luan kai zengzhi de kouzi*), and they even blindly engage in undesirable competitions with each other (*mangmu huxiang pengbi*). They lavish in extravagance, use their limited resources to buy cars, they blindly put money into show-off projects, take on capital construction, and fiddle with construction projects where they shouldn't. This not only increased their financial difficulties, but also bred corrupt behaviour. What could we do? The central government's macroeconomic policies were not ours to meddle in, so we could only take the road of increasing control over township-level fiscal administration.[470]

In order to achieve this goal, the county authorities applied a Weberian approach to economic management vis-à-vis its seven townships and fourteen towns. The underlying principle was that township-level administration should be improved, standardized and subjected to better supervision. Its fiscal rights, however, should not be touched. This principle was handily summarized as 'the six that will not change' (*liu ge bu bian*). Specifically, the reform should not change the distribution of benefits at the county and township levels (*xian xiang liyi fenpei*), the township level's right to make budgetary allocations (*xiangzhen de yusuan fenpeiquan*), the property and usage rights over township-level finances, (*xiangzhen zijin de suoyouquan he shiyongquan*), the right to examine and approve fiscal transactions (*caiwu shenpiquan*), the independent accounting units (*duli hesuan zhuti*), and the relationship between creditor's rights and liabilities (*zhaiquan zhaiwu guanxi*). All township-level revenue would first be passed upwards, and finances then reallocated according to expenditure needs (*shouru shang jiao, zhichu xia bo*).

Three small groups of five accountants each were formed to annul and collect 1,115 tax and 25,649 fee blank receipts and to close 248 township- and village-level accounts. Thereafter, receipts would be handed out 'in cooperation' with the county government (*piaoju tongguan*). 27.7 million yuan withheld from funds designed for the payment of construction costs and wage arrears were confiscated. Instead, standardized accounts for fiscal expenditures, wages, settling balances and village finances were opened in each township and town, but they were co-administered by the county (*zhanghu tongshe*). A summary was made of present liabilities and creditor's rights, and a plan drawn up to reduce township- and village-level debts. The grassroots administrations were 'in principle' forbidden to incur new liabilities, and exceptions had to be granted by the township-level 'fiscal management group', which was headed by the Party secretary and the magistrate, and approved by the county finance bureau. Resources for non-payroll personnel were withdrawn.[471]

After this preparatory phase, measures were implemented to standardize budgetary procedures, strengthen administration, and improve the transparency and supervision of township-level fiscal administration. Specifically, the division between employees financed from budgetary revenues and those financed from off budgetary revenues was suspended, and all wages now had to be paid out of the budgetary accounts. Expenditure quotas were drawn up for wages, administration costs, hosting delegations, conducting meetings, and buying cars. Expenditures for local development and for attracting capital had to be individually applied for. In addition, standardized and formal rules for handling accounts, receipts, fiscal reserves and public acquisitions were drawn up. Administration was to be strengthened by first separating the core duties of education and birth control from other day-to-day work and putting them under special administration. Second, the county was to staff the township-level governments' finance offices, and take responsibility for their training, supervision, and evaluation, while the township-level government remained in charge of the day-to-day administration. Since most fiscal planning was now conducted at the county level, the post of the township-level 'general budget accountant' (*zong yusuan kuaiji*) was abandoned. County- and township-level finances were to be administered together by means of a shared software and intranet connection. For the village level, only one 'village-level collective finance account' (*cunji jiti zijin zhuanhu*) administered also by the county was opened, and village-level tax collection and social service provision were to be centralized in 'agricultural tax collection parlours' (*nongyeshui nashui dating*). Finally, transparency and supervision were to be strengthened by implementing a responsibility system and ordering supervisory organizations to show 'no tolerance' (*guxi*) towards individuals and organizations who violated the new rules.[472]

According to Zhou Weihan, the reform was well received at the township level, because 'salaries were now all paid on time and in full amount. In addition, finances for relieve measures now reached their addressees, which improved the relationship between peasants and cadres.' In addition, the workload of township-level cadres had been reduced, since the former accountants could now engage in the still tedious work of collecting and administering the agricultural taxes. Finally, township-level expenditures in 2003 were reduced 12 per cent over the previous year, and liabilities 8.7 per cent. At the same time, budgetary revenues increased 20 per cent, because 'the reform motivated the townships and towns to tap their full revenue potential'.[473] However, Zhou also stressed that 'the fiscal base of township-level governments is still very fragile, and their fiscal difficulties have not yet been fundamentally remedied'. In particular, there was additional need to 'analyse the relationship between administrative behaviour, administrative costs and administrative achievements', since the problem of township leaders diverting funds into show-off projects had not yet been completely solved.[474]

These were only two, albeit important, examples of complementary reform pioneers. Many more existed: on March 19, 2003, Anhui provincial government honoured 178 'advanced units' and 274 'advanced individuals' for their achievements in Rural Tax and Fee Reform pilot work. The above paragraphs have shown that the claimed achievements are credible, but that even where the county leadership threw their full weight behind the experiments, they were not easy. Where such commitment lacked, reform implementation was far more difficult.

Resisters

It is safe to say that few counties across China were as committed to making the Rural Tax and Fee Reform work out as Taihe, Wuhe or He County. Yet even in those counties where such extraordinary zeal was present had it been difficult to carry the reform into the townships and villages. Zhou Weihan's argument that the Complementary Reform measures also benefited township-level cadres bears some validity, yet it should not blind us to the fact that they first and foremost reduced the power, authority and resources of local leaders. It is easy to comprehend that many grassroots leaders resisted the changes envisioned by the higher levels. Three resistance strategies with different degrees of legality can be distinguished: continued illegal fee-taking; exploiting legal or systemic loopholes to extract revenue from subordinate governments and, ultimately, the peasants; and reducing rural service provision.

First of all, fee-taking continued in the very places where peasant income needed to be raised most. Several 'Urgent Notifications' demanding cessation of such practices were passed down by all administrative levels in 2001 (see above). According to a member of Anhui's RTFR Leading Small Group, an 'enormous number' of peasants came to Hefei, the capital of Anhui province, to petition (*shangfang*) against illegal fees.[475] Obviously, illegal levying abated only after the agricultural tax was totally rescinded in 2006. Interviews conducted in two villages of Hebei's P County in summer 2004 and in one village of Anhui's G County in spring 2006 confirm this. Although peasants in both Hebei villages confirmed that all fees had been abandoned and that taxes were now collected only twice a year, they were given neither advance notifications nor receipts. In addition, they did know that an agricultural tax reform had been taking place, but they had no idea how the new tax was calculated and how much they had to pay. As one peasant remarked, 'they just come to your house and ask for a certain sum of money. You have no choice but to pay them. Maybe they charged me too much, maybe they didn't. I don't know.'[476] The relationship between peasants and cadres was very tense. Only two weeks before I arrived in that particular village, one respondent had his door broken open by a gang of township cadres: 'My wife could not show up for the birth control check-up because she works in the city, so we had to pay a fine. When we refused, they came with two cars, broke open the door, and took our DVD player. Then they were gone again'.[477] A young peasant from the other village also told of strained relations between the villagers and the township cadres. He had owned a restaurant, but saw himself forced to close it down because his main customers, delegations from the township, never paid for their meals:

> Every time they gave me an IOU and told me to get my money from the township government. When I went there, they said it was none of their business. Finally, they owed me more than five thousand yuan. Every year at Chinese New Year I go there to ask for my money, and every time they only give me five hundred yuan. This way, it will take me ten years until the account is settled.[478]

Besides continuing illegal practices, local governments engaged in semi-legal and legal ways of extracting peasant revenue. A study of four provinces (Jiangsu, Anhui, Heilongjiang, Guizhou) by three researchers of the Agricultural Ministry's Office of Rural Reform Experimental Districs (*Nongyebu nongcun gaige shiyanqu bangongshi*) illustrates how unclear regulations were exploited by county cadres for this purpose. For example, villagers were made to pay not only for roads built within a village, but also for roads connecting villages and even townships. In many places, peasants working in factories were required to pay for their absence from the village. 'Voluntary donations' by parents to their children's schools had replaced the now illegal study fees.[479] The new regulations which allowed townships to manage village funds and counties to manage township funds did not improve the situation. On the contrary, they paved the way for county governments handing down unrealistic, though perfectly legal achievement targets for the collection of the agricultural tax. Besides illegal debts, austerity measures and belated payment of government employees' salaries, the only way to generate revenue in areas without a significant secondary sector were the agricultural taxes and their surcharges. As said before, the fulfilment of tax duties was a major item in the evaluation of local-level cadres. Not handing up the contracted amount in taxes resulted in the deduction of the missing amount from the responsible government's budget and sometimes even from the salary of the responsible cadres. This made tax collection an issue of primary importance, especially so at the township level. Because of this, village cadres were given no choice but to increase the peasant burden. Before the agricultural tax was scrapped, the 'tax enforcement effect' (*bi shui xiaoying*)[480] was a common occurrence: village cadres pushed peasants to pay their tax arrears, if they had money or not. This was not illegal before a central-level document prohibited the practice in 2002 (see above),[481] yet was made responsible for the suicide of many indebted farmers.[482] The ultimate abolition of the agricultural tax was most probably due to its inherent unfairness and the low level of burden reduction achieved, especially for those peasants who needed it the most.

The third strategy to consolidate grassroots government revenue was the reduced provision of government services. As one county chief put it quite drastically: 'Basically, we have stopped building roads. . . . We would do anything but fire people. How should we decide who to fire? Everyone in the township line offices has relationships with someone at the county level. This would lead to a major conflict'.[483] Systemic factors hindering personnel reductions existed as well, especially in those provinces where the counties had not taken over township-level finances. As Caijing journalist Chang Hongxiao points out with respect to reform implementation in Heilongjiang province,

> project-specific funds and the projects themselves are controlled by the higher-level governmental organizations. If a township or town does not have a vertical branch of that organization, it will encounter all sorts of restrictions. These restrictions often directly manifest themselves in very specific losses, for example the inability to receive certain funds.[484]

In other words, budgetary restraints of varying firmness were imposed on township-level governments, but incentives to establish additional organizational units and to hire personnel out of extra-budgetary revenues persisted. Similarly, higher-level delegations still expected to be wined and dined, and not everywhere were show-off projects and ineffective capital building discouraged. Several researchers point out how rural infrastructure projects suffered from the reforms, widening the gap between countryside and city even further. Examples in case include the building of roads, the sanitation of school buildings so dilapidated they are declared 'dangerous', and the maintenance of waterways and dams.[485] Education, one area specifically targeted for improvement, suffered especially. As Thomas Bernstein illustrates, the problem of unpaid teacher salaries worsened drastically in Anhui province during the first reform year. As a consequence, primary and middle schools charged students in excess of the level set by the central government, a practice that was formally condoned by at least one county government. Many teachers left their jobs, which resulted in overcrowded classes and eventually forced some students to quit school.[486] In some places, the situation not even improved after the counties had become responsible for financing rural compulsory education. One village school I visited in central Anhui's G County was so dilapidated that the headmaster forbade me to take pictures. He feared that he could be reprimanded if they were published. The faculty were very unhappy and complained that they were owed six months' wages, and that they had never ever received full payment. When they complained, they were told that 'the county does not have the money.'[487] In addition, many townships were forced to plunge themselves further into debt. As the head of the RTFR group of Anhui's W County made clear, the county leadership fully condoned this. Again, concrete figures are hard to come by, but the development of township-level debts in He County provides a good example. There, township-level liabilities increased from 67.7 million yuan in 2000, the first year of the reform, to 81.0 million yuan only one year later. In 2002, township-level debts had increased to 87.7 million yuan.[488]

A final strategy that has been mentioned several times throughout this study and which probably best fits into the semi-legal type is false reporting. As the author's interviews with township officials from Wuhe County revealed, the reduction figures cited above need to be treated with care. According to the leader of the Organization Department, the number of payroll positions was indeed reduced, but by far not as substantively as the above figures suggest.[489] A cadre dispatched from the city Fiscal Bureau to work in the township confirmed this and added that the same applied to non-payroll personnel (*chaobian renyuan*). 'They pose a much larger burden on the diminished township finances than the payroll personnel, as the township receives no funds from above to remunerate them. In order to pay them, the government must sell public property like real estate and abandoned schools', the official confided. Due to scarcity of such resources, this cannot be a long term strategy; the local cadres, however, see no alternative. When asked what the township will do when all the property is sold, the official answered 'we'll see then' (*dao shihou zai shuo*).[490] This illustrates a well known problem and merits scepticism regarding the sustainability of these measures: a job on the government payroll is the only means of employment in many underdeveloped regions.[491] As State Council Development Research Centre researcher Zhao Shukai put it:[492]

> Social stability is now a major constraint for structural reforms in townships and towns. One situation that governments at every level do not want to incur is a petition army of township and town cadres joining with the already existing petition armies of the unemployed and the peasants.

A final point that needs to be made is that it was not only grassroots cadres that applied resistance strategies, but also the peasants. The Party secretary of a village in G County pointed out to the author that newspaper reports about peasants' enthusiasm for the new tax scheme were not true, at least not for his village:

> Tax collection was very cumbersome. I myself had to walk from door to door to collect them. When we were not allowed anymore to collect tax arrears, many villagers said: 'If they don't have to pay, I won't pay either'. They felt that the system was unjust, rewarding those who refused payment. Therefore, we were able to collect only about 70 per cent of the tax quota in 2002, 2003 and 2004, and that was already very much. The neighbouring village only managed to collect 50 per cent.[493]

This does not seem to have been an isolated incident. In February 2004, before the abolition of the agricultural tax was proclaimed, Anhui Province ceased collection of the supplement to the agricultural tax that had replaced the village retention. According to the document, 'the peasants . . . do not understand and accept the supplement; its administration is rather complicated, causing much work and high costs. Furthermore, due to various reasons, it is very difficult to collect the full sum, . . . so it cannot guarantee the coverage of village-level expenditures. Finally, . . . burden is still rather high, especially for those peasants who farm a comparatively large acreage of contracted land'.[494]

The Party secretary also pointed out that the system of 'one undertaking, one decision', where public construction projects are democratically decided and financed, was basically sabotaged by the villagers:

> Take street building, for example. If we needed money to build a street, we had to walk from house to house to solicit opinions. This is difficult, because so many work outside. Then they ask if the street will run by their house. If it does not, they refuse to take part. However, in reality 100 per cent must be willing to give some money. If 80 per cent are willing, and 20 per cent are not, the 20 per cent cannot be forced to pay. Accordingly, the other 80 per cent do not want to pay either.[495]

He complained that his responsibilities remained, but that he had no way to ask the peasants for money:

> Yesterday, an inspection group from the township came. It would be a loss of face not to wine and dine them. But there are no more funds for this sort of thing. What should I do? I had to pay the meal out of my own pockets. Or take the electricity cables. They were stolen several times. It costs 5,000–6,000 yuan to replace them, but there are no funds for that. I had to pay them myself.[496]

The villagers confirmed this, but also added that he could afford it: the Party secretary had much land, and he had the biggest house in the village. However, they hastened to add that he was wealthy before he became Party secretary, and that he was not corrupt.[497] When asked why he did this sort of job, he replied that 'it gives me face'.[498] The villagers also confirmed that they were not willing to participate in financing any public projects. They enjoyed being freed from taxes and fees, and they relished that 'nobody can come now and demand money from us'.[499] Since many of them worked in the nearby city, they did not see the necessity of improving their village. When the new central policy of connecting all villages by a paved road was being realized, the township's plan was to sell the trees that lined the dirt track in order to pay for the road. This was also refused by most villagers.[500]

Initial results

On the surface, the reform appeared quite successful. Officially, Anhui had already achieved a 37.5 per cent burden reduction following the first reform stage,[501] and burden reduction in all twenty provincial-level entities that had implemented the reforms averaged 45.8 per cent.[502] As these figures do not account for the illegal fees and levies that had supposedly been abandoned, the percentage should be much higher. Second, according to Minister of Finance Xiang Huaicheng, the reform had achieved a substantive simplification of Party and government administrative structures at the township level, at least in Anhui Province. Whereas, on average, townships had five to eight Party and government organizational units before the reform, the number was reduced to one to three. Altogether, 9,700 units were dissolved. In addition, township-level units were decreased from 1,841 to 1,760, and village committees from 30,658 to 29,186 between 2000 and 2002, a reduction of 4.4 per cent and 4.8 per cent, respectively. By these measures, personnel on the administrative payroll were reduced 10 per cent and service payroll personnel even 50 per cent. In total, 40.6 per cent of 'people eating the emperor's grain' were reduced.[503] The 'successes' regarding other complementary measures were not elaborated in such detail. Basically, the formulaic prose of the previous documents was repeated, with the difference that 'great steps' towards fulfilling the goals formulated therein had ostensibly been made. As one county-level cadre stated it in another context, 'this is a sure-tell sign that they have discussed these issues at great length, but nothing else. If they had something to boast with, would they not do that?'[504] The same might be true for prefecture-level reports for this reform stage. An internal documentation made available to the author which contains the 'achievement sheets' of all of Anhui's prefecture-level governments illustrates this well. While, for example, Suzhou and Fuyang proudly list their impressive administrative, service and village-level personnel reductions, Huainan merely states that

> In line with the spirit of document . . . [we] actively and reliably engaged in the reform of the organizational structure of townships and towns. Holding fast to the necessities of the Socialist Market Economy and rural social development, [we] resolutely transformed government functions, improved the work style, and strengthened administration of and services to society according to the principles of Seeking the Truth from Facts, simplification and efficacy, combining responsibility and authority, and Administration According to Law. On the basis of the basic principles of township organizational reform and political necessities, we strictly applied measures such as diversification, merger and change according to local circumstances. Township-level organization units with similar functions, overlapping tasks, and rather unitary work responsibilities were simplified and adjusted. After simplification

and unification, the previously six offices of each township and town organ were in general reduced to only two, and the township-level service units were generally reduced to only six.[505]

Obviously, the township-level administrative units themselves were not reduced, but only the number of offices within them. Unfortunately, the reader is denied the quantitative impact of this rigorous implementation of central and provincial policy. The only figure supplied, i.e. the reduction of village-level cadres, is quite low in comparison even if the rather small administrative size of Huainan City is considered. Table 6.1 lists all the figures given in the internal reports, calculated into percentages.

Although the information presented here is far from complete, Table 6.1 clearly shows that substantive variation underlies the aggregate achievements. This is generally not only true for aggregate provincial-, but also for city-level data. The last chapter has shown that counties have chosen their individual ways of implementing national polices, and the next chapter will make clear that this results in outcomes that are very different across counties even within a prefecture. Why is that? Besides differences in economic circumstances, political factors figure strongly. As has been the

Table 6.1 Reduction of administrative and service units, personnel, village cadres, 2000–2001 (%)

Village	*Administrative units*	*Service units*	*Administrative personnel*	*Service personnel*	*Village cadres (persons/village)*[a]
Hefei	−66.6	−55.4	−10.0	−25.7	3.1
Huaibei	−67.7	−40.0	−10.0	−22.4	1.0
Bozhou	n.a.	n.a.	n.a.	n.a.	n.a.
Suzhou	−77.6	−50.0	−10.0	−51.5	n.a.
Bengbu	−70.5	−56.5	n.a.	n.a.	−1.6
Fuyang	n.a.	n.a.	−15.4	−74.4	1.2
Huainan	n.a.	n.a.	n.a.	n.a.	0.2
Chuzhou	n.a.	−42.9	−10.0	n.a.	1.2
Laian	n.a.	n.a.	−5.8	−37.2	1.3
Maanshan	n.a.	n.a.	n.a.	n.a.	n.a.
Chaohu	−82.4	−46.7	n.a.	−26.2	2.0
Wuhu	n.a.	n.a.	n.a.	−9.0	1.8
Xuancheng	n.a.	n.a.	n.a.	n.a.	2.0
Tongling	−66.7	−64.3	n.a.	n.a.	n.a.
Chizhou	n.a.	n.a.	−3.0	−2.6	n.a.
Anqing	n.a.	−43.4	−10.0	−21.5	n.a.
Huangshan	n.a.	n.a.	n.a.	n.a.	n.a.
Average	−71.9	−50.0	−9.3	−30.1	−0.7

Source: Own calculations from Anhui Province RTFR LSG, *Ge shi nongcun shuifei gaige shidian gongzuo zongshu*, internal reference, 2003.

Notes: [a] Since only absolute figures for village personnel reduction were provided, I divided the number of reduced cadres by the number of villages for each prefecture-level government in 2002. *n.a.* Not available.

case with all Chinese reforms so far, successful policy design and implementation hinged on the role of pioneering localities who manage to transfer abstract requirements into workable policies. Other localities, in contrast, resist central-level orders because they are unwilling, and often unable, to implement them.

Persisting problems

Not only do the results look so spectacular when they are disaggregated, but problems continued to surface. While the reform had fermented in Anhui province for more than four years and had experienced various adaptations to local circumstances, the other provinces were not involved in this process and had to implement a prefabricated policy. Especially at the grass roots, there was not much support for the RTFR. Obviously, the combination of subsidies, competition under hierarchy and a heavy hand was not enough to rein in the resistance.

Less than half a year after the reform extension, three central-level circulars (two of them 'Urgent Notifications') indicated that the reform was not being smoothly implemented. On August 20, an 'Urgent Notification' complained that local leaders did not take the reform seriously.[506] Furthermore, central-level subsidies frequently did not reach their destination and were even subject to embezzlement. Therefore, the document urged to implement a 'responsibility system' that made leading cadres accountable for 'burden-related ugly occurrences' (*shjie nongmin fudan de exing anjian*) and increasing petitions. On October 24, the Minister of Finance held a lengthy speech at a RTFR symposium in which he praised the achievements of the RTFR.[507] According to Xiang, 'this year, there have clearly been less visits and complaints (*shangfang*) by farmers than before the reform' in the major agricultural regions Anhui, Henan and Hunan. Several sections on, however, he complained that 'since recently, there has been an increase in letters and complaints by farmers in the pilot regions, the majority of which reflected that implementation at the grass roots was not up to standard, and that there were policy deviations and problems in the cadres' work style'. Two weeks later, the General Office of National Tax Administration followed up on Xiang's speech and admonished tax collectors to follow existing collection guidelines.[508] They were especially to refrain from 'using their tax collection authority to collect add-on fees [*da che shoufei*] . . . to violate tax collection policies, [and] to levy the tax on agricultural special products per capita'. A 'minority' of tax collectors and assistants were accused of 'lacking warm feelings and responsibility towards the farmers, of having a lax attitude towards policies and the law, of having a simple and rough work style (*gongzuo zuofeng jiandan cubao*), and even of illegally confiscating peasant property.'

Hence, the tax administration formulated the 'ten not-alloweds' (*shi bu zhun*), which give a good overview over the resistance strategies discussed before. Tax collectors were specifically not allowed to

1. Violate tax and administrative laws and regulations and increase the scope of collection.
2. Collect in excess, in advance, or to apportion taxes.
3. Illegally grant tax rebates.
4. Not hand out receipts or to hand out vouchers.
5. Establish tax collection contracts.
6. Illegally use force.
7. Employ the police force or police instruments.
8. Arbitrarily arrest people, confiscate food, goods, and animals.
9. Hand out irregular fines.
10. Use tax collecting power for levying other fees.

One week later, another urgent notification alerted cadres that

> since 2002, several places have engaged in following up on peasants' tax and fee arrears in the context of implementing the RTFR. Due to the fact that the origin of these arrears has complex causes, clearing these arrears can easily lead to new contradictions and difficulties. There are some places where the fiscal capacity of farmers is not taken into account in the process of settling tax and fee arrears, and where arrears are forcefully cleared. There are some places where the scope of collection is arbitrarily widened, and where arrears from wanton levying, wanton fund-raising and wanton apportionment are also figured in. There are places where cadres conduct the collection in a simple fashion and with a rough work style, and they even form hit squads that grab produce and take away belongings. This has led to extreme dissatisfaction on the side of the peasants. In individual places (*gebie difang*), ugly occurrences and mass incidents occurred where people were hounded to death (*bisi renming*). This has caused extremely adverse (*elie*) effects in society and has become a major hidden danger (*yi da yinhuan*) to the healthy development of the Rural Tax and Fee Reform and to rural stability.[509]

Again, personal responsibility and drastic measures against violators were promoted. Despite these obvious difficulties, the reform was announced a success and became national policy on March 27, 2003. The experiences made during this period seem not to have validated the usefulness of the policy. Less than one year later, on March 5, 2004, Premier Wen Jiabao declared that the agricultural tax would be 'gradually phased out' within five years, which amounted to a radical policy change.[510] The new rural tax system was basically being abandoned. Another year later, in the Government Work Report to the third session of the tenth National People's Congress in March 2005, Wen announced that the agricultural tax would be completely rescinded by 2006, two years ahead of schedule.[511] Already in 2005, twenty-six province-level governments had scrapped the tax, leaving only the provinces of Hebei, Shandong, Yunnan and Gansu charging a reduced rate of 2 per cent for one more year.[512]

Abolition of the agricultural tax

Basically, the years 2004 to 2006 were a period of transition from the original reform design, which was based on competition under hierarchy, to more purely centralized and hierarchical central–local relations. The central-level document that heralded the gradual abolition of the agricultural tax in 2004 justified this step with the statement that 'not all problems in the RTFR pilot regions have been solved. The peasant burden is still rather heavy, and complementary reform implementation lags behind. The fundament of peasant burden reduction is not yet consolidated.'[513] Another indicator for the failure of the original reform design, besides the continued application of the resistance strategies outlined above, is that the centre started to further centralize rural reform work. Instead of pushing the localities to think up ways to improve rural development, it took the matter into its own hands. Two months after the aforementioned document was promulgated, another State Council circular approached all relevant ministries to form inter-ministerial 'specialized small groups for drafting guiding documents to deepen the work in the RTFR pilot regions' (*shenhua nongcun shuifei gaige shidian gongzuo zhidaoxing wenjian zhuanti xiaozu*). In particular, ten areas were to be covered by one group each: township-level structural reforms; the administration of rural compulsory education; rural finances; rural public hygiene; grassroots-level liabilities; rural tax and fee arrears; 'one undertaking; one decision' and rural public services; state-owned farms; the overall tax system; and legal reform.[514]

This transition was reflected in a change of names. The Rural Tax and Fee Reform was officially ceased in 2006 and seamlessly gave way to the Rural Comprehensive Reform (*nongcun zonghe gaige*) which was part of the 'New Socialist Countryside' programme.[515] On August 24, 2006, the RTFR groups and offices were accordingly renamed by order of the State Council.[516]

On the local level, this transition was reflected in a widening of focus from merely pressuring local governments to implement austerity measures to a centrally guided and–financed development programme. Premier Wen made clear that the accomplishment of the Main Reform, the scrapping of the Agricultural and related taxes has to be followed by a period of consolidation. The focus would be on the reform of grassroots governments, their fiscal systems and improving rural education.[517] Anhui provincial government was equally ambitious, starting to clear up the enormous amount of rural liabilities in order to 'further reduce the debt burden of county and township-level governments' and 'completely re-energize county and township finances' within 'three to five years.'[518] To further this aim, the Centre allocated an additional 15 billion yuan RMB for a programme called 'three rewards and one subsidy', which rewarded counties and townships that increased their revenue and streamlined their administration. Superordinate governments were rewarded for subsidies to 'counties in financial straits' and for successful debt reduction.[519]

Since late 2003, pressure mounted to implement administrative reduction measures also outside Anhui. Among others, Hubei, Liaoning, and Jiangsu provinces have started to implement township-level personnel cuts on a trial basis. They ranged from one half to two thirds in selected counties or county-level cities, and radically restructured township-level administrations. According to official figures, in Hubei's trial townships and towns, for example, the Political Consultative Conference was abolished, and each Party Committee was reduced to a maximum of nine persons, with the Party secretary concurrently serving as head of government. Township-level offices were reduced from fifteen to three, service provision thus reduced to (primary and middle) schools, hospitals and finance offices. The number of persons on the government payroll was cut to thirty-seven to forty-five from about 100. In 2004, also according to official figures, Jiangsu province reduced the number of township-level employees by 40,000, the targeted figure being 100,000. Of the 17,000 service units scheduled for reduction, 7,000 had already been merged with other units.[520] In late 2004, equally impressive results were reported for Hubei's Gan'an district, Jianli county, Honghu City and Macheng City (see Table 6.2).

Very likely these are only isolated examples, and as the last sections have shown, they need to be treated with care. Recently, the central government imposed a five-year moratorium on township-level employment, which signifies a bouncing back of personnel numbers in recent years.[521] Significant strides had only been made with the merging of townships and villages in individual provinces and counties, which had already reached its apex in 2001, one year before the reform was extended. Again implementation was very uneven. An overall reduction of 20.7 per cent of township-level governments and 13.1 per cent of villages between 2000 and 2005 was, for example, matched by a 38.2 per cent reduction in the number of township-level governments

Table 6.2 Reduction of personnel and administrative entities in four county-level governments, Hubei province, 2004 (%)

Entity	*Administrative entities*	*Government and Party group personnel*	*Administrative employees*
Gan'an District	−68.7	−61.2	−71
Jianli County	−40	−38	−44
Honghu City	−40	−51	−35
Macheng City	n.a.	−48	−59

Source: Y. Li, 'Hubei Sheng xiangzhen zonghe peitao gaige dui nongcun hexie shehui goujian de jiji yingxiang', *Guojia Xingzheng Xueyuan*, 2005. Online. Available HTTP: www.nsa.gov.cn/download/liyoucai.doc (accessed on September 23, 2005).

in Ningxia, but only of 0.6 per cent in Hebei.[522] It has to be added that significantly more townships than towns were reduced. This is an indicator not only of recentralization by means reducing the number of territorial units, but also of China's urbanization strategy.

Changes in the (township/town) administrative structure, however, remained cosmetic outside (and often inside) Anhui, and personnel were transferred rather than reduced.[523] Anhui had the idea to combine personnel reduction and local development, so it put 9,000 former township cadres into village branch secretary positions.[524] Strictly speaking, this was illegal, as the branch secretary has to be elected by the village Party members. The aforementioned report states that the practice of staffing paid village positions with township cadres was adopted by other provinces as well.[525] At the same time, local governments were supposed to increase spending on rural compulsory education and social security. Spending on these two sectors was to grow faster than overall government expenditures. The difficulties in this transition were aggravated by the gradual abolition of the agricultural tax, which affected local revenues severely in 2004, 2005 and 2006.

Fortunately, the centre's increasing fiscal commitment not only offset, but exceeded these losses (Table 6.3), even if they did not cover the total revenue reduction since 2002, which had added up to 125 billion yuan RMB.[526]

While the centre had been willing to invest only 20 billion yuan in the Rural Tax and Fee Reform in 2000, this sum almost quadrupled within only six years. In 2006, the centre had transferred 78 billion yuan in RTFR earmarked subsidies, and in mid-2007 Minister of Finance Chen Xiwen

Table 6.3 RTFR earmarked transfer payments from the centre and revenue reductions from abolition of agricultural tax, 2000–2006

Year	*RTFR subsidies (billion yuan RMB)*	*Increase over previous year (billion yuan RMB)*	*AT revenue reduction (billion yuan RMB)*	*Difference between reduction and subsidies (%)*
2000[a]	1.1	–	–	–
2001[a]	1.7	–	–	–
2002	24.5	–	–	–
2003	30.5	+6.0	–	–
2004	52.4	+21.9	−18.6	+3.3
2005	66.2	+13.8	−18.2	−4.4
2006	78.0	+11.8	−5.9	+5.9
2004–06	–	+47.5	−42.7	+4.8

Sources: S. Zhu, H. Zhang and H. Yan, 'Nongcun shuifei gaige shidian he xiangcun guanli tizhi gaige genzong yanjiu baogao', *Jingji Yanjiu Cankao* 40, 2003, pp. 2–24 ; M. Su, *The Major Innovations regarding the Support of the 'Three Rurals' with National Finances: an exegesis of the 2006 fiscal budget report*, Beijing: Ministry of Finance, 2006. Online. Available HTTP: http://www.mof.gov.cn/news/20060310_2264_13666.htm (accessed December 1, 2007).

Note: [a] Anhui Province only.

announced that subsidies for 2007 would increase even more.[527] And this was not all. In 2004, central-level spending on rural development increased 22.5 per cent over the previous year to 262.6 billion yuan. In 2005 and 2006, subsidies increased another 13.3 and 14.1 per cent, respectively. Transfer payments for direct grain subsidies and rural compulsory education marked the largest increases, although the latter was primarily targeted at western China. Direct grain subsidies increased 130 per cent between 2005 and 2006 (27.6 billion yuan), and another 63 per cent in 2007 (42.7 billion yuan). Also included in this sum is increased spending for providing access to drinking water (6 billion yuan) and electricity (2.5 billion yuan), improving rural infrastructure (43 billion yuan), and a new scheme to provide minimum living subsidies in the rural areas. Very importantly, all these subsidies were earmarked, so they could only be spent for their intended purpose.[528] Obviously, the centre realized that its combination of administrative delegation and competition under hierarchy had failed. As a consequence, administrative rights and responsibilities were centralized, and the steering of local policy implementation reverted to more purely hierarchical patterns.

Conclusion

This examination of the RTFR's implementation process has provided important insights regarding not only the impact of China's structure and processes of central–local relations on policy outcomes, but also the relationship between policy design and implementation. As the previous chapters have shown, the Anhui provincial and county-level governments were the main arenas where agenda-setting took place. Conversely, the formulation of the official RTFR policy was rather a top-down affair. In contrast, the implementation of the reform again depended on the cooperation and ingenuity of the counties and especially the township-level governments. As this chapter has made clear, the provinces relayed downwards the task of adapting the centre's abstract demands to 'local circumstances', and eventually this task fell to the counties. Competition under hierarchy put great stress on them, but at the same time gave them some autonomy to fill the abstract policies with meaning. Some took up the challenge and came up with meaningful reform proposals, while others chose to resist because they either lacked the means to do so or had from the start not supported this policy. In either case, the pressure fell on the townships that were forced to increase their performance while at the same time cutting their costs.

Contrary to the initial reform design, the counties were not made subject to institutional reforms and austerity measures. Since the higher levels depended on their ingenuity in contextualizing or merely implementing the reform, the county level emerged strengthened. Given the narrow corridor in which local innovation could take place, however, it is more appropriate to speak of delegation than of devolution. Hence, the counties were given limited decision-making powers within a corridor constrained by the centre's achievement standards. For the township-level governments, the situation was very different. No matter if their county government resisted the policy or cooperated with the higher levels in bringing the RTFR to fruition, they had to do its bidding. In this context, two factors merit special attention. First, it has been pointed out that the relationship between counties and townships is far more centralized than intergovernmental relationships at the higher levels. In addition, while the governments above the county level apply a mixture of hierarchical orders and the stimulation of competition to steer policy implementation, the county level solely relies on hierarchical steering vis-à-vis the townships and towns. Second, it was shown that initially, burden reduction hinged on the local governments' implementation of austerity measures that would free the necessary funds for this venture. Again, the grassroots administrations were the sole targets of these measures, and it is no wonder that many shirked their implementation since they were faced with various conflicting incentives.

The reform did not change the fact that the township-level governments were chiefly responsible for the provision of local and grassroots social services. In addition, they were still required to help fulfil the county governments' tax quotas. It is no wonder that townships that had no communal enterprises to help them accomplish this difficult feat needed to rely on credits and exactions from the peasants besides scaling back the provision of social services. Coupled with the inherently unjust design of the new agricultural tax, these exactions eroded the peasants' support for the new policy, which further diminished revenue and increased the pressure at the township level to generate the necessary resources to meet the quotas specified by their super-ordinate counties. Hence, the reform clearly discriminated against the poorer townships which did not have the resources to shoulder the reform costs.

As was shown, these difficulties had become obvious already in the first reform year, but the central government did not address them properly. More regulations mirroring earlier circulars were passed, and they were underscored by more pressure on local-level leading cadres to implement them properly. It must be added, however, that the central government was no unitary actor in this respect. While the Ministry of Finance applied a hard line vis-à-vis the local level, Premier Zhu Rongji had good reasons to sympathize with the plight of the grassroots governments. The failure of the Highway Fee reform in 1998 and the Grain Purchase Regulation's clash with the rural fiscal reform experiments had already given observers of China's politics the impression that Zhu was not very versed in rural affairs. His widely publicized discussion with township leaders and rural primary school teachers in Anhui clearly confirmed this observation, which was increasingly being shared by China's general public. More than ever did he need the RTFR to succeed, so he was more ready than the Minister of Finance to pay the necessary price. This confirms Linda Li's finding that the reform implementation was marred by factional strive, bureaucratic cellularization and a lack of 'expertise and interest' in the relevant departments.[529] However, the increased transfer payments, many of which did not reach their destination, were not enough to ensure the cooperation of the grassroots governments. In addition, they did not address the inherent unfairness of the new agricultural tax and solve the problem of declining tax receipts. As a consequence, the reform had to be aborted, and the new agricultural tax was quickly rescinded.

It follows from these observations that the initial reform design inherently discriminated against those groups which were supposed to benefit from it: the poor. Since poor townships predominantly relied on agriculture for revenue, burden reduction not only affected their revenue base, but concurrently their ability to offset these losses with non-agricultural revenue. As for the peasants, the reform benefited rich and poor peasants alike, and did not change the regressive character of rural taxation. Still, poor peasants had to submit a substantively larger part of their revenue to the local fiscal offices than rich ones. If this is true, then the reform would not have had an equalizing effect, but would indeed have widened the gap between rich and poor both with respect to local government revenue and peasant income. The next chapter will test these hypotheses by providing a quantitative analysis of the reform outcomes in Anhui's sixty-one counties.

7 Impacts and outcomes: the case of Anhui Province

As the preceding chapter has shown, the RTFR did not proceed as had been initially planned. Various problems surfaced during the implementation process, compelling the central government to abandon the initial policy in early 2004. The nature of these problems was laid out in detail, and it was indicated that the impact of the reform at the various stages of the implementation process was far from uniform across Anhui's sixty-one counties, 1,455 townships and towns, 24,076 villages, and 13,461,300 households.

This chapter aims to shed some light on to these differences. Focusing on the distributive effects of the reforms at both the peasant and the county government levels in China's Anhui province, it seeks to deepen our understanding of the reform process and its outcomes. On a methodological basis, quantitative and geographic methods serve to augment the overwhelmingly qualitative analysis presented in the previous chapters. The level of analysis will be the county level, mainly for the reason that comprehensive data for the township and village level are not available. Having stated that the townships suffered most from the impacts of the RTFR, and bearing in mind that even the townships within one county display widely differing levels of development, such an analysis might at first seem unsatisfactory. These limitations are acknowledged, but there are many reasons in favour of such an analysis. First and foremost, the county level's considerable autonomy and its status as the most important local-level development agency has been pointed out already in the introduction. Second and relatedly, the level of affectedness of a county government determines its ability to level out negative impacts of the reform on townships, villages, and even households.

The first part of this chapter is descriptive and outlines the aggregate changes in the peasant burden, peasant per capita income as well as government receipts and expenditures during all phases of the reform. As these sections will make clear, the peasant burden has decreased and peasant income increased, but considerable variation underlies these changes. The same is true for government finances. However, such descriptive statistics do not provide proof that these changes were a direct or even indirect result of the RTFR. Indeed, loss of income and revenue might be the result of external factors as diverse as China's WTO accession in 2001, the severe flood that swept across the country in 2003, or other policies implemented concurrently. Conversely, increasing peasant per capita income (PPCI) might not be a result of the reform, but of increased migrant labour. In order to clarify these issues, the second part of this chapter applies multivariate statistics to data for Anhui's sixty-one counties. As a first step, it examines the structural determinants of peasant income and government revenue one year before the reform started. As can be expected, the size of the industry sector is found to be the single most potent variable determining aggregate both peasant per capita income and government revenue. It then analyses the impact of the RTFR on peasant income and government finance. Specifically, it tests the following hypotheses: first, the overall growth of peasant per capita income is not the result of the RTFR, but of the growth of the industrial sector as well as migration. Second, the RTFR has had adverse effects: it made poor governments even poorer. Since this loss in revenues was partly compensated by an increase in government transfers, a third hypothesis pertains to the use of these transfer payments: we expect them to be earmarked, so the third hypothesis to be tested is that transfer payments were used to finance social services and local development.

Aggregate development of peasant income and government finance

As Chapter 5 has shown, the aims of the Main Reform were the reduction of the peasant burden and the increase of peasant income. At the same time, the Three Guarantees stipulated that service provision should continue, especially in the education sector. The following section explores how the peasant burden, peasant income, government revenue and government expenditures developed in the course of the reform. Specifically, it provides data for the pre-reform year 1999, the year preceding the decision to scrap the agricultural tax (2003), and the year the tax was abandoned everywhere in Anhui (2006).

As Table 7.1 shows, the reform has significantly reduced the peasant burden in Anhui's sixty-one counties already in the first reform phase. Exactions from Anhui's farmers fell from an average of 110 yuan per person to 59 yuan, a decrease of around 47 per cent. However, burden reduction was far from uniform across Anhui province. While per capita peasant burden in Dangshan County was reduced 75.6 per cent during the first stage of reform, the corresponding figure for Dangtu County was only 3.9 per cent. The examination of peasant income at these three points in time yields a similar picture. Average per capita peasant income grew steadily from an average of 1972 yuan per person in 1999 to 2,087 yuan four years later, and to 3,104 yuan in 2006. As for revenues and expenditures, what strikes the analyst first is that total per capita revenue decreased between 1999 and 2006 (after a sharp drop in 2003), but that per capita budgetary expenditures more than doubled during the same time period (Table 7.2).

Table 7.1 Peasant per capita burden and income in Anhui's sixty-one counties, 1999–2006

	Burden per capita		*Peasant PCI*		
Measure	*1999*	*2003*	*1999*	*2003*	*2006*
Mean	109.90	58.90	1971.80	2087.20	3104.20
Minimum	67.40	32.00	1104.00	1211.10	1724.67
Maximum	163.30	116.90	2548.00	2964.00	5065.93

Sources: Anhui provincial government, RTFR LSG, *Anhui Sheng nongcun shuifeigaige qianhou nongmin fudan qingkuang biao*, internal reference, 2003; *Anhui tongji nianjian*, Hefei: Anhui Tongji Chubanshe, various years.

Table 7.2 Budgetary revenue and expenditure in Anhui's sixty-one counties, 1999–2006

	Rev2/Cap	*Budget Rev/Cap*		*Budget Exp/Cap*		
Measure	*1999**	*2003*	*2006*	*1999*	*2003*	*2006*
Mean	243.1	182.0	299.4	304.5	441.4	904.7
Minimum	144.6	64.2	58.2	184.8	208.1	407.4
Maximum	446.6	498.0	1,289.6	531.4	1,016.7	2,212.6

Sources: Anhui provincial government, RTFR LSG, *Anhui Sheng nongcun shuifeigaige qianhou nongmin fudan qingkuang biao*, *Anhui tongji nianjian*, various years.

Note: For the year 1999 the important yet non-budgetary receipts from the village levy, the township comprehensive fee and the education fund are added to that year's budgetary revenue. This adjusted revenue will be referred to as Rev2 throughout this chapter. The same method will be applied to correct expenditure (Exp2). The peasant burden for 1999 will be calculated by calculating the percentage of all agriculture-related regular taxes and fees in Rev2. Since all non-budgetary levies were abandoned in 2000, such adjustments are not necessary for the following years. It suffices to relate the receipts from agricultural taxes, which have all become budgetary revenue, to total budgetary revenue of that year.

In fact, what characterizes the reform period is that the ratio of county revenues to expenditures increased rather steeply since 1998. Whereas in 1998, Anhui county-level expenditures on average exceeded revenues by around 27 per cent, the ratio more than tripled within only seven years! This suggests that during this period of time, revenue inequality has increased, and that attempts have been made to address this problem by means of fiscal transfers. This issue must be explored further. Figure 7.1 illustrates the development of the Gini-coefficients for aggregate county-level revenue, expenditures as well as PPCI. The Gini coefficient is a measure of inequality, with a higher coefficient signalling higher inequality.

Figure 7.1 confirms that revenue inequality between Anhui's counties grew to alarming proportions in the course of the reform. It also shows that

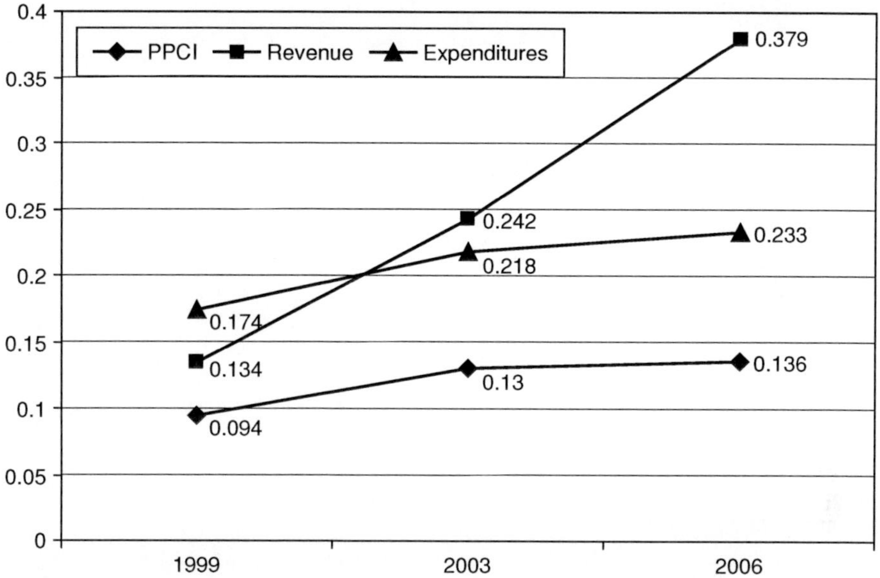

Figure 7.1 Gini coefficients of peasant per capita income, budgetary revenue and expenditures, Anhui County level, 1999–2006

in the first reform stage, higher revenue inequality translated into higher expenditure inequality. This provides first evidence for the hypothesis that the reform had disequalizing effects on county finance in Anhui. Strangely, rising inequality can also be observed for peasant incomes during the first reform stage. It is interesting to note, however, that both peasant income and county expenditure inequality were mitigated in the second reform stage. Figure 7.2 proves that the lesser incline of expenditure inequality is indeed the result of targeted fiscal transfer payments by the central and provincial governments. It visualizes the relationship between government per capita revenue and transfer payments in 2005 and contrasts it to the situation in 1999.

As can be seen, the size of transfer payments only very weakly correlated with government per capita revenue in 1999. However, the determinants of transfer payments changed dramatically in the course of the reform. While in 1999, the variation in per capita revenue explained only 9 per cent of the variation in transfer payments (r^2 = 0.09), the two variables largely corresponded six years later. In 2005, the size of per capita revenue explained 66.1 per cent of the variation in transfer payments (r^2 = 0.661). In simple terms, they now clearly came to benefit those counties where government per capita revenue was low. Since these, as will be seen, also tend to be the counties that were dominated by agriculture and where peasants earned comparatively little, they can be considered structural subsidies.

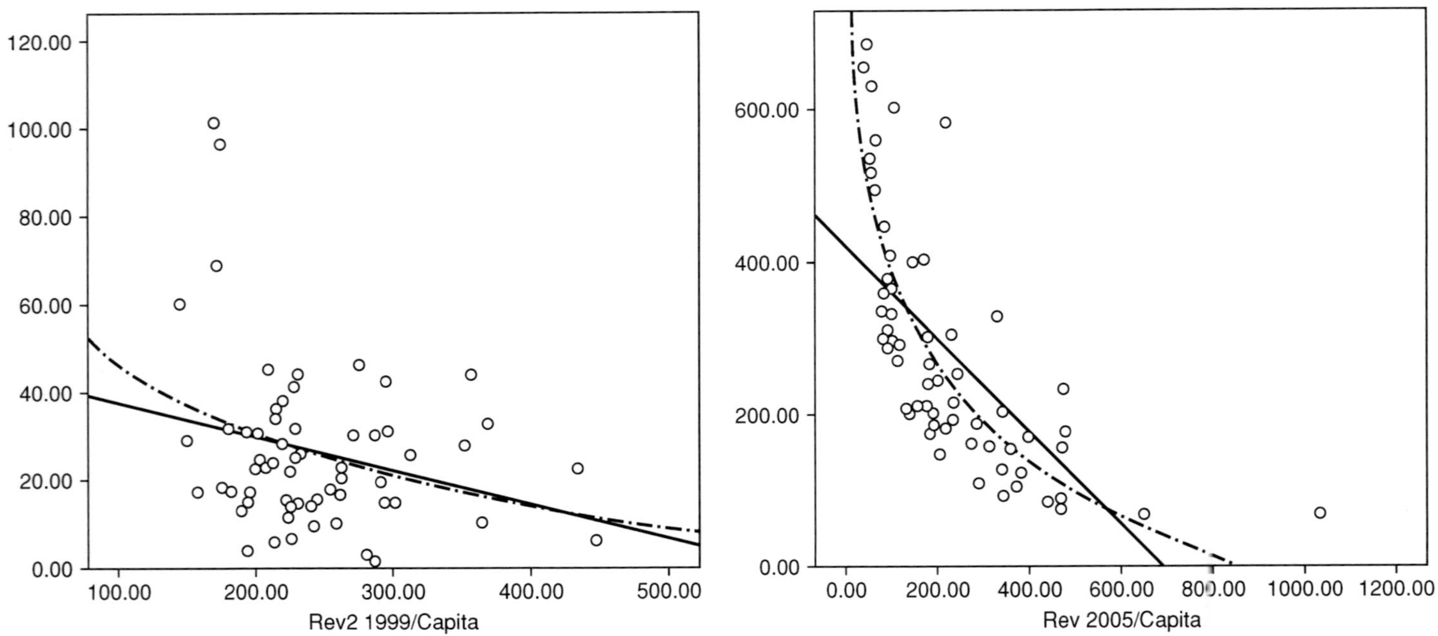

Figure 7.2 Relationship between per capita county revenue and transfer payments, 1999 and 2005

The structural determinants of peasant income and government revenue

What was responsible for these changes? The following sections seek to provide explanations that account for the differences just outlined. In particular, they test three alternative hypotheses. First, the differences in development are a function of structural factors that have preceded the reform. In other words, unequal development has been path-dependent and was unrelated either to the RTFR or to other policies. Second, the differences were brought about by the reform policies. Third, they were brought about by factors neither related to the underlying economic and social structure, nor by reform policies, but by other factors that have yet to be identified. In order to test these hypotheses, we will start out by analysing which factors determined PPCI and government revenue in Anhui Province one year before the reform started.

First, a decision must be made on how to operationalize such 'structural' factors, and what other variables to include. For example, a higher PPCI in eastern Anhui might have been caused by increased economies of scale in agriculture, by peasants migrating to the wealthy neighbouring provinces, or by improved opportunities to engage in sideline occupations. Other variables might be responsible as well, such as socioeconomic and socio-structural factors, involvement in other sectors than agriculture, the social infrastrucure, or natural disasters. However, running a linear regression that includes all of these indicators would prove fruitless for two reasons: first, including such a large number of indicators violates the principle of parsimony, which calls for explaining as much as possible with a model as simple as possible. Second, and more important, some of these indicators might be intercorrelated which would skew the outcome and possibly render it meaningless.

In order to reach the twin goal of providing a parsimonious explanation and of not excluding any indicators with explanatory value, we will run a factor analysis including various indicators for the economic structure;[530] for the level of socioeconomic development;[531] and more general development indicators.[532] Table 7.3 summarizes the result of the analysis for the pre-reform year 1999.

The factor analysis reveals that there indeed exist significant correlations between the considered variables. As can be seen, only two components account for about 62 per cent of the whole set's variation. Factor 1 alone explains 37 per cent of this variation, and as Table 7.3 shows, this factor is related particularly closely with the local economic structure. It correlates very highly with all indicators for non-agricultural economic development in general, and with industrialization in particular. Interestingly, this factor correlates negatively with the relative size of the primary sector, indicating that a high level of industrialization and a strong agriculture tended to be mutually exclusive before the reform commenced. As the analysis reveals, this factor also correlates significantly with development indicators such as access to

Table 7.3 Main components extracted from various development indicators for Anhui's sixty-one counties, 1999

	Component			*Component*	
Indicator	*1*	*2*	*Indicator*	*1*	*2*
GDP/Cap99	0.944	0.011	Electricity/RuralPop99	0.366	−0.143
Secondary/Cap99	0.917	−0.073	Land/RuralPop99	0.055	0.920
Tertiary/Cap99	0.898	−0.168	Sown/RuralPop99	0.119	0.901
Nonagripop/TotalPop99	0.834	0.114	Wage/TotalRuralInc00	−0.125	−0.682
Secondary/TotalGDP99	0.803	−0.198	AgMach/RuralPop99	0.305	0.677
Primary/TotalGDP99	−0.801	0.424	Primary/Cap99	0.520	0.564
TotalPop99	−0.624	0.341	Tertiary/GDP99	0.169	−0.526
DrinkingWater/RuralPop99	0.575	0.364	RuralHouseholdSize99	−0.501	0.510

Source: Analysis based on *Anhui tongji nianjian*, 2000.

Note: Extraction method: main component analysis.

drinking water, the share of the population not engaged in agriculture, rural household size, and electricity use per capita. This illustrates the large impact of a viable industrial sector for China's development, and it simultaneously illustrates that the size of the tertiary sector did not play a determining role in this respect. Since Factor 2 correlates very highly with land per person and the sown acreage per person, and also with the amount of agricultural machinery employed per peasant and the primary GDP per person, economies of scale in agriculture most likely underlie the second factor. Very interestingly, the relative size of wage receipts in peasant per capita income correlates negatively with this factor, suggesting that peasants in agriculture-intensive regions were not likely to seek or obtain employment elsewhere. Thus, the initial seventeen structural variables have been reduced to merely two, namely non-agricultural economic development and economies of scale in agriculture.

How well can these and other, non-structural factors explain variations in peasant income and government revenue? Table 7.4 provides the Pearson correlation coefficients for the bivariate correlation of the two factors and the these variables in 1999. As for what determines the size of PPCI, the explanatory value of two other important non-structural variables, namely the peasant burden and natural disasters (percentage of total sown area that was affected by flood and drought) is also tested. In the case of government revenue, we are not only interested in the impact of industrialization and scale economies. In addition, we test two propositions made earlier: first, poorer peasants tend to live in poorer counties; and second, it is mainly the poor governments that depend on the peasant burden for revenue.

As expected, PPCI in 1999 correlates most significantly with an industrialized economy. Equally interesting, neither is the average peasant burden level statistically related to this variable. In other words, high average peasant per capita burden did not automatically lead to low average PPCI, and vice versa. This is a very significant finding, since it has frequently been assumed that a high peasant burden tended to lead to the stagnation of peasant income.

Table 7.4 Determinants of PPCI and government revenue in Anhui's sixty-one counties, 1999 (Pearson correlation coefficients)

	Non-agricult. economy	*Agric. econ. of scale*	*FloodDrought/ TotalSown99*	*PeasantBurden/ Cap*
PPCI	0.638**	0.214	−0.349**	0.063
	Non-agricult. economy	*Agric. econ. of scale*	*PPCI*	*PeasantBurden/ Revenue*
Rev299/Cap	0.762**	0.109	0.560**	−0.780**

Sources: Analysis based on *Anhui tongji nianjian* 2004 and *Anhui tongji nianjian* 2006.

Notes: **The correlation is significant at a level of 0.01 (two-sided). *The correlation is significant at a level of 0.05 (two-sided).

Thus, the hypothesis put forward in Chapter 4 is verified: it is income inequality which made the peasant burden a problem, and not the amount of fiscal exactions per se. Since the county-level aggregate figures employed here are averages and do not reflect this inequality, they do not correlate with the size of the peasant burden. As might be expected, however, the percentage of sown crops destroyed by flood and drought correlates negatively with peasant per capita income.

As for total government revenues in 1999 (including agricultural levies and fees), the level of industrial development is identified as the major determining variable as well. Clearly confirmed is the hypothesis that the poorer a county, the lower average PPCI tends to be. Since almost 90 per cent of the population at the county level and below are registered as 'agricultural population' in the Statistical Yearbooks, this is not surprising. Peasants are the chief source of government receipts at this level, and high peasant income is reflected in higher government revenues.

The hypothesis that poorer governments tend to rely more on the peasant burden for revenue is confirmed even more clearly. This confirms the observation made by Thomas Bernstein and Xiaobo Lü that local economic development has been the major intervening variable determining the impact of efforts to reduce the peasant burden.[533] Prosperous provincial-level east-coast (*dongbu*) governments like Shanghai, Guangdong, or Jiangsu with a sound non-agricultural revenue base could easily bear the costs of the reforms, while most counties in the so-called central (*zhongbu*) provinces depended mainly on exactions from the peasants for revenue.

Explaining the changes in peasant income and government finances

Having established the determinants of PPCI and government revenue before the reform commenced, the three hypotheses regarding the impact of

Table 7.5 Determinants of PPCI change in Anhui's sixty-one counties, 1999–2003 and 2003–2005 (Pearson's *R*)

Burden/Cap	*Primary GDP/Cap*	*Secondary GDP/Cap*	*Tertiary GDP/Cap*	*WageInc Cap*	*HouseholdInc/ Cap*
			1999–03		
0.431**	0.737**	0.310*	−0.044	0.126	0.281*
			2003–05		
0.067	0.482**	−0.049	0.088	0.592**	0.407**

Sources: Analysis based on Anhui provincial government, RTFR LSG, *Anhui Sheng nongcun shuifeigaige qianhou nongmin fudan qingkuang biao; Anhui tongji nianjian*, various years.

Notes: **The correlation is significant at a level of 0.01 (two-sided). *The correlation is significant at a level of 0.05 (two-sided).

the RTFR on PPCI and government finance can now be tested. Since the reform proceeded in two stages, the outcomes of the two stages and their determinants will be compared, rather than merely contrasting the *status quo ante* and the *status quo post.* In particular, such a comparison might yield further explanations for why the initial reform design was abandoned.

To test the first hypothesis that the overall growth of peasant per capita income is not the result of the RTFR, but of the growth of the industrial sector as well as income from migration, three sets of indicators are examined for their impact on PPCI change between 1999 and 2003, and between 2003 and 2005: The direct impact of the RTFR (measured by the decrease of peasant per capita burden); economic growth in the primary, secondary and tertiary sectors (measured by the increase in GDP per capita in each sector); and the increase of wage and non-wage income.

The partial hypothesis that structural factors are responsible the most important determinant of PPCI growth between 1999 and 2003 is verified. Interestingly, it was not the growth of the secondary, but of the primary sector that most strongly affected PPCI in the first stage of reform. The explanation for this might be that while the peasants are the main producers of the agricultural GDP, only a part of them is employed in industry. As for the impact of the RTFR on PPCI, there is, contrary to what we expected, a clear relationship between burden increase and PPCI growth. However, the relationship is positive, meaning that there was a slight tendency for PPCI to increase at a lower rate where more burden was reduced. Obviously, there must be one or more intervening variables which had a positive impact on burden reduction, but a negative impact on PPCI growth. The hypothesis that the RTFR was not responsible for PPCI growth in the first reform stage is nevertheless verified.

The same holds true for the second reform stage from 2003 to 2005, albeit with an important modification: the increase in wage income became the single most important determinant of the increase in PPCI. This confirms an internal report of Anhui's Statistical Bureau, which states that migrant labour income is responsible for 85 per cent of the increase in peasant income.

On the other hand, there is a significant correlation between the increases in PPCI and non-wage income as well, which at first sight might be due to the fact that peasants have received subsides for planting non-cash crops since 2004. As Heerink, Kuiper and Shi have shown, however, this was not the case. Rather, the positive development in market prices for food grains is the single most potent factor that has induced the peasants to return to agriculture. The comparatively little subsidies were found to not have had any impact in this respect.[534]

The second hypothesis concerned the adverse effects of the RTFR on government revenue: it stated that the reform contributed to making poor governments even poorer. Again, the alternative hypothesis that economic development was responsible for government revenues will be considered. In addition, the development of peasant gross income, the primary tax source base until the second reform phase set in, will also be included into the analysis.

Table 7.6 clearly illustrates the devastating effects of the RTFR on county revenue in both reform phases: The more a county depended on the peasant burden for revenue, the more drastic were its revenue losses. As for the economic factors, in the first reform phase there are some similarities to the impacts observed with respect to PPCI, which is easy to understand: as shown above, peasant incomes grew with the agricultural GDP, and PPCI and government revenue are closely correlated. Thus, the RTFR as well as economic factors had an impact on government revenue. In other words, in the first phase of the reform those counties fared worst who not only

Table 7.6 Determinants of change in per capita government revenue in Anhui's sixty-one counties, 1999–2003 and 2003–2005 (Pearson's *R*)

		1999–03		
Burd/Rev(2)	*Primary GDP/Cap*	*Secondary GDP/Cap*	*Tertiary GDP/Cap*	*PPCI*
−0.715**	0.430**	0.360**	−0.205	0.642**
		2003–05		
Burden/Rev 2003	*Primary GDP/Cap*	*Secondary GDP/Cap*	*Tertiary GDP/Cap*	*PPCI*
−0.767**	−0.145	0.290*	−0.530**	0.200

Sources: Analysis based on Anhui provincial government, RTFR LSG, *Anhui Sheng nongcun shuifeigaige qianhou nongmin fudan qingkuang biao: Anhui tongji nianjian*, various years.

Notes: **The correlation is significant at a level of 0.01 (two-sided). *The correlation is significant at a level of 0.05 (two-sided).

depended on the peasant burden for revenue, but who at the same time suffered from a stagnating agricultural sector.

Figure 7.1 visualizes the deleterious effect that the reductions had had on government receipts in most counties. Of ten counties with revenue increases, only five registered gains above 10 per cent. Four of them belonged to the rather industrialized Yangze abutters Maanshan and Wuhu City, and the other was coal-rich Fengtai County. Losses were especially grave in northern Anhui, where burden had been heaviest and revenue lowest.

As can be seen, the losses are most significant in poor north-western Anhui. Not incidentally, eight of the most affected counties were listed among the least developed by almost all indicators listed in Anhui's Statistical Yearbook 2006, and four of them are even counted as 'national level poor counties'.[535]

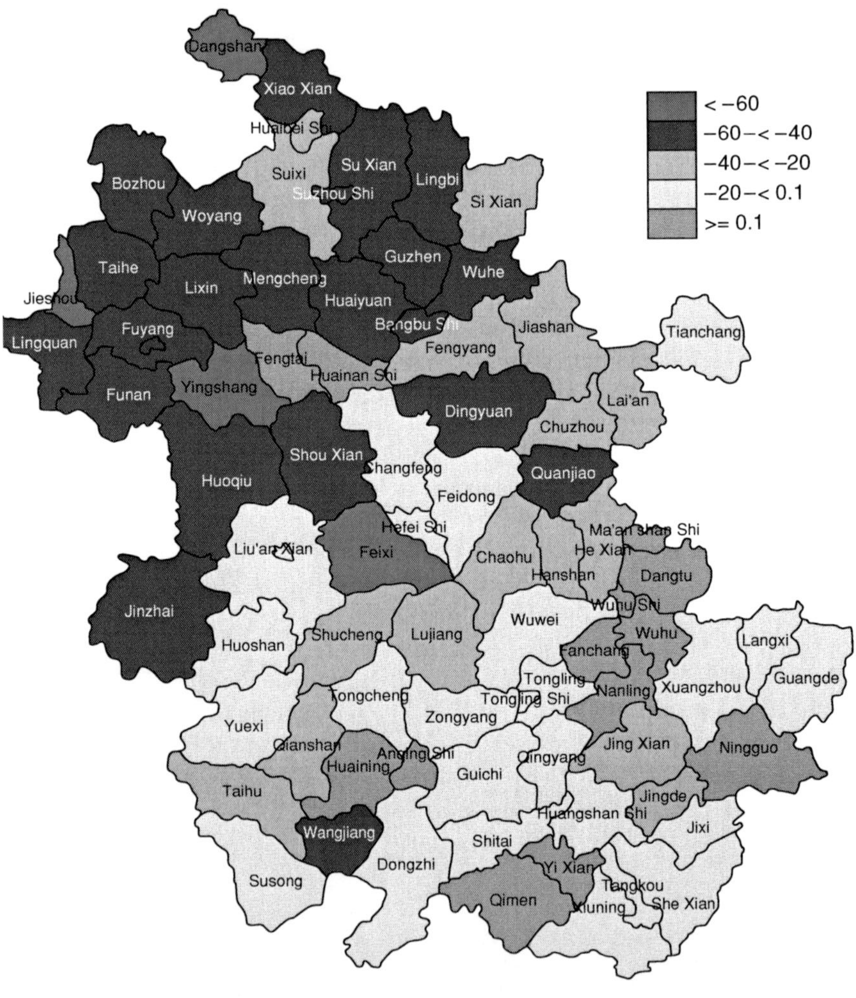

Map Spatial distribution of per capita budgetary revenue decrease, county level, Anhui Province, 1999–2003

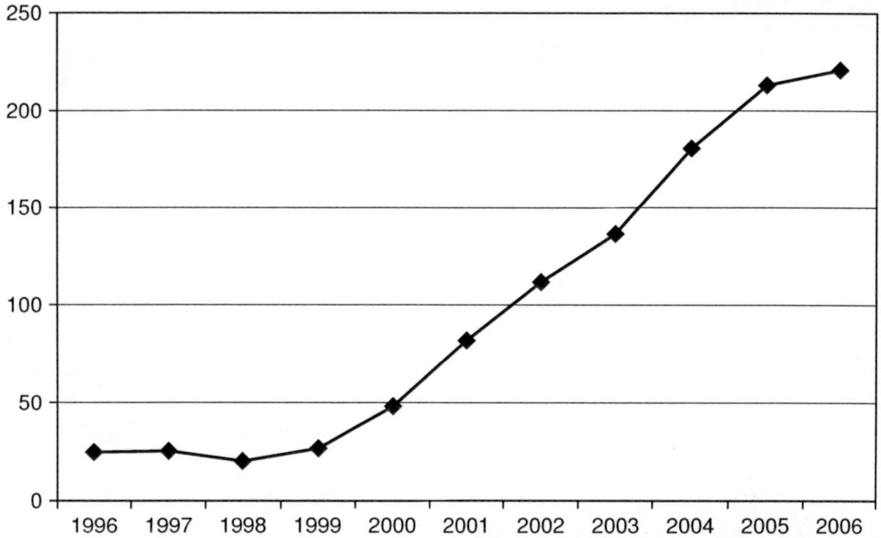

Figure 7.3 Percentage of budgetary expenditures exceeding budgetary revenues, Anhui Province, county level, 1996–2006

In the second phase, the economic indicators surprisingly loose their explanatory power, but high dependence on the peasant burden is the single most potent variable. It alone explains almost 60 per cent of the variation in per capita government revenue between 2003 and 2005. This takes us to the third hypothesis, which holds that increased transfer payments to the counties were used to finance social services and local development. To recall, while expenditure developments closely mirrored revenue trends during the first phase, a regularized fiscal balancing mechanism was gradually being put into place from 2003 on. Figure 7.3 illustrates the widening gap between revenues and expenditures, especially after the year 2000.

This adds weight to the observation made in the previous chapter, namely that county and grassroots governments resisted implementation of the original reform design and successfully engaged in a tug-of-war with the central government. Because transfer payments increased so steeply from 2001 on, administrative restructuring and personnel reduction measures could be further resisted.

After the initial policy had failed, the local governments had to be compensated for the loss of revenue resulting from the gradual abolition of the agricultural tax. In 2003, one year before the agricultural tax was gradually abolished, the tax made up 34 per cent of county-level revenue, so the centre's involvement had to become even larger. This is illustrated by the large leaps in the expenditure–revenue ratio in 2004 and 2005. As a consequence, all counties were able to steeply increase expenditures despite declining revenues especially during the second reform phase.

Expenditures

Given that the centre's policy focus has partly shifted from economic growth to rural development, and given that a large part of the transfer payments are directly or indirectly financed by the centre, we would expect the expenditure structure to mirror this shift in foci. In general, the share of expenditures for agricultural development, education, health and social security should increase faster than those for other, non-rural development oriented items. Furthermore, we would expect the alleged success in personnel reduction to be reflected in decreasing or at least stagnant administrative expenditures. It has been shown above that revenue-poor counties benefited most from the transfer payments. As, not incidentally, revenue poverty correlated highly with average peasant poverty, we would expect the shift in expenditures to be even stronger there. In order to identify general trends in policy priorities, we first examine how expenditure structures on the aggregate city and county levels have developed throughout the RTFR period. Thereafter, we test how far revenue increases in general, and transfer payments in particular, have affected county-level spending on social services.

The following sections trace the priority changes in public service provision during the RTFR period. Unfortunately, it is beyond the scope of this book to gauge the effects of the reform on public service provision itself. Therefore, I will mainly focus on the important question if spending patterns have changed during the reform period, as this is an important indicator for the commitment of local governments to improve rural development.

As Figure 7.4 shows, in terms of their relevance in total budgetary expenditures at the city level education spending has unexpectedly remained rather stable in Anhui Province. At the county level, which has to shoulder the overwhelming part of education expenditures in China, the ratio is higher. The growth trend, however, is very similar (2000: 29.2 per cent; 2002: 31.5 per cent; 2005: 27.0 per cent). Obviously, there has been no strong shift in priorities for this item. The other social service-related expenditure items have also remained quite stable over time. The only exception are subsidies to social security programmes, which have taken a steep incline between 1999 and 2002, and a less marked decline since then. Remarkably, aggregate city-level expenditures on health, support for underdeveloped areas, pensions and agriculture are far below the expenditures for retirement pensions for administrative personnel. Similarly, the provincial government spends only slightly more on subsidies for the social security programmes for millions of people than it does on pensions for former civil servants. It should be noted, however, while the expenditure share for government administration has steadily declined both at the city and the county level, absolute spending increased rather steeply (18.2 per cent between 2000 and 2001 alone, and 30.0 per cent between 2003 and 2004).

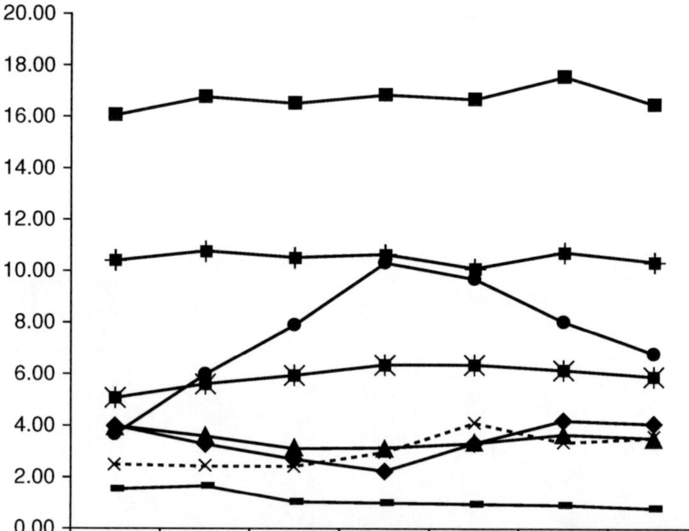

	1999	2000	2001	2002	2003	2004	2005
Agriculture	3.98	3.30	2.69	2.23	3.33	4.21	4.06
Education	16.05	16.74	16.51	16.83	16.66	17.55	16.47
Public Health	3.95	3.63	3.16	3.13	3.37	3.68	3.51
Social Welfare	2.49	2.46	2.45	2.93	4.10	3.36	3.59
Retired Admin	5.06	5.63	5.93	6.38	6.36	6.15	5.92
SocSec Subsidies	3.66	5.97	7.89	10.31	9.68	8.04	6.79
Administration	10.38	10.78	10.51	10.64	10.11	10.72	10.35
Underdev. Areas	1.53	1.65	1.03	1.00	0.95	0.89	0.81

Figure 7.4 Expenditures on various social services as a percentage of total expenditures, Anhui Province, aggregate city level, 1999–2005

As had been the case with peasant income, burden reduction, and government finances, Anhui's sixty-one counties again display a rather wide variation with respect to their spending on education, social security and health. As Figure 7.5 indicates, relative spending on education decreased in the majority of counties during the first phase of the reform, and it decreased further between 2003 and 2005. Outlays on social security, on the other hand, grew steadily in almost all of the counties (Figure 7.6). Impressive as this growth might at first seem, it needs to be kept in mind that the outlays for social security are minor and that therefore even a relatively small increase in absolute terms is reflected in a rather large growth rate. Health spending remained rather consistent in relative terms, but again the interregional variance must be considered rather high, with Dongzhi County spending six times more of its budget on health than Mengcheng County (Figure 7.7).[536]

It follows from these observations that, with the possible exception of the social security sector, the general increase in expenditures during the two phases

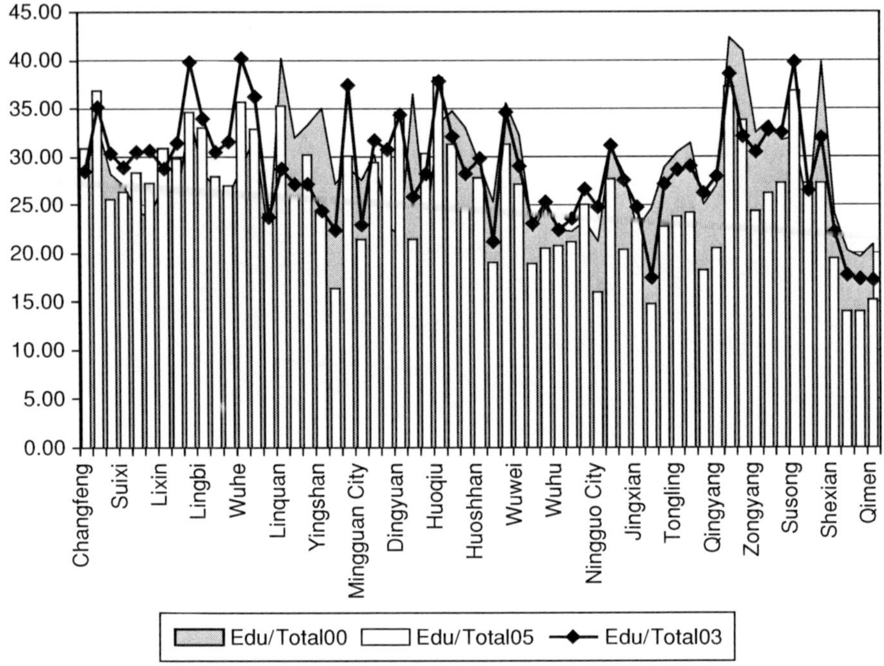

Figure 7.5 Expenditures on education as a percentage of total expenditures, Anhui Province, county level, 1999–2005

of the RTFR has not been specifically targeted at improving the public service infrastructure. The statistical analysis confirms this and also shows that transfer payments were obviously not tied to the improvement of the public service infrastructure. As a general rule, the increase of expenditures for all the items discussed so far followed the general increase in expenditures. None of them has been especially strengthened, at least not during the first phase (Table 7.7).

Table 7.7 contrasts the change in expenditure structures in the two phases[537] with the corresponding increases in budgetary spending and in transfer payments. In the first phase, the slight changes in expenditure priorities observed above cannot be explained by either the amount of expenditures or the size of transfer payments. The only exception is education, where expenditures tended to increase at a lower rate then total expenditures. In addition, there was a slight tendency that counties which received higher transfer payments spent less on public security. In the second phase, the picture somewhat changes. Expenditure-rich counties display a neglect of spending on agriculture and social security, although the correlation is quite weak. However, they tend to spend more on items that are not individually listed in the yearbook. As for those counties who benefited from large transfer payments in 2005, they display a tendency, albeit a slight one, to

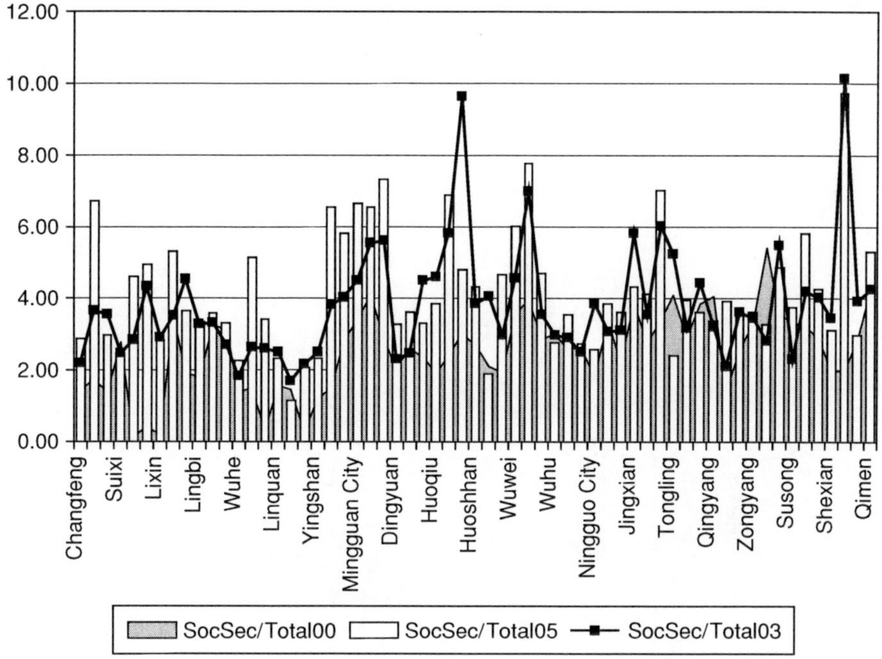

Figure 7.6 Expenditures on social security as a percentage of total expenditures, Anhui Province, county level, 1999–2005

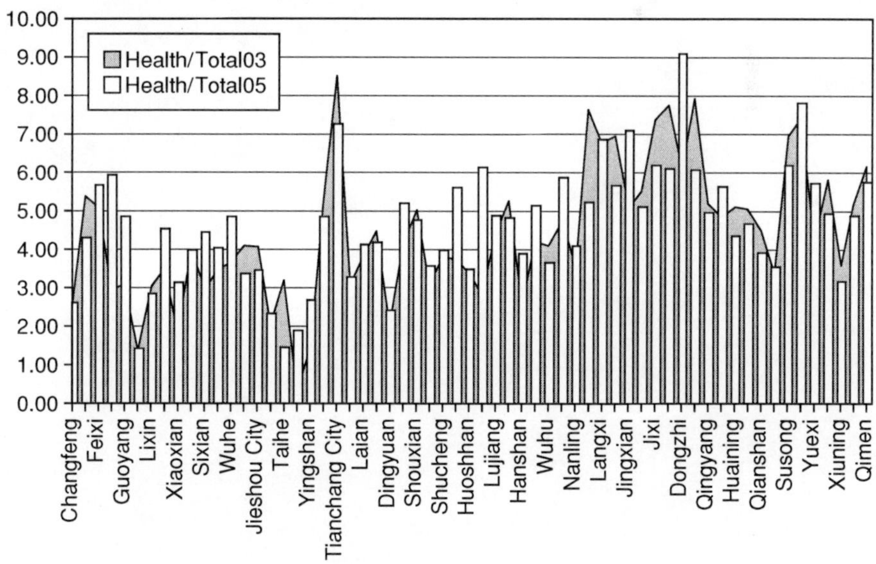

Figure 7.7 Expenditures on health as a percentage of total expenditures, Anhui Province, county level, 2003–2005

Table 7.7 Determinants of public service expenditures for Anhui's sixty-one counties, 2000–2003 and 2003–2005 (Pearson correlation coefficients)

Determinant	*CapitalConstr.*	*Agri*	*Edu*	*SocSec*	*Admin*	*PubSec*	*Other*	*Health*
				2000–03				
Exp/Cap 00–03	−0.008	0.108	−0.440**	0.017	0.078	−0.125	0.041	n.a.
Rev/Exp 2003	−0.015	0.057	−0.133	−0.057	−0.213	−0.364**	−0.064	n.a.
				2003–05				
Exp/Cap 03–05	−0.140	−0.297*	−0.216	−0.259*	−0.045	−0.217	0.338**	0.162
RevExp 2005	0.269*	0.101	0.358**	−0.112	0.036	−0.079	−0.134	0.286*

Sources: Analysis based on Anhui provincial government, RTFR LSG, *Anhui Sheng nongcun shuifeigaige qianhou nongmin fudan qingkuang biao*; *Anhui tongji nianjian*, various years.

Notes: **The correlation is significant at a level of 0.01 (two-sided). *The correlation is significant at a level of 0.05 (two-sided) focus more on education and health.[538] To conclude, the assumptions put forward at the beginning of this section have not been proved true. With the possible exception of education, where there was a slight tendency for counties with large transfer payments to increase relative spending, the central government's focus on rural development was not reflected in changing expenditure structures. Again, this does not mean that nothing happened. First, expenditures on items related to social services tended to increase along with the general increase in expenditures, so additional money was definitely being spent. Second, as Brandt et al. point out, much investment in rural infrastructure was being made via earmarked transfers from the centre that do not appear in the regular budget.

Conclusion

Summing up the observations made in the previous sections, the following picture emerges. On the aggregate level, the RTFR had no impact at all on aggregate county-level PPCI. Of course, this does not amount to stating that the reform did not achieve anything in this respect. It is very likely that poor peasants did indeed profit from the reduction in regular and irregular exactions. As Chen Xiwen has shown, the regressive nature of the peasant burden affected the poorest peasant strata the most, which had an average of more than 37 per cent of its gross revenue reduced by taxes and fees.[539] For the richest strata, such deductions only amounted to less than 5 per cent. What the data do imply, however, is that the reduction of income inequality achieved by the reform was surely outstripped by the increase in PPCI in Anhui's more prosperous regions. In other words, on the aggregate county level, it was not the RTFR that made peasant incomes increase, but economic factors such as a growing primary sector in wealthy regions and income from non-agricultural activities.

As for the development of fiscal revenues, the analysis has shown that two non-interrelated factors were chiefly responsible: first, the increase in peasant income and second, burden reductions. Low PPCI tended to cause low revenues, and these revenues were even further reduced by the RTFR and the tax adjustment reform. Hence, the initially poor counties became even poorer, while the wealthier counties were able to shoulder the losses incurred by the reform by mobilizing other revenue. Geographically, this led to the amplification of Anhui's south–north divide. Since the initial reform design had not foreseen transfer payments and basically demanded that counties implement austerity policies to offset the losses, the northern counties were even more disadvantaged. Their reluctance and probably also inability to implement such measures is probably one of the chief reasons while the initial reform design had failed. Fortunately, the abolition of the agricultural tax forced the central government to address this situation, and a mechanism of transfer payments was institutionalized that benefited revenue-poor counties over their wealthier counterparts. In fact, every single county spent more in 2005 than it had earned, but the margin underlying the size of transfer payments was huge.

Remarkably, the centre has not utilized its increased financial involvement towards adjusting the expenditure structure of the subsidized counties. While comparatively higher spending on items related to the provision of social services would have been expected especially in those counties that profited most from the transfer payments, this assumption was falsified. Either this change in focus was operationalized outside the regular budget by means of earmarked transfers, as Brandt et al. have found out, or other factors related to the structure of central–local relations had been at work.

8 Why the 'Third Rural Revolution' failed

This study has set out with the overarching assumption that the institutionalized structures and processes of central–local relations were responsible for the various unintended outcomes of the RTFR. Specifically, this study threw some light into the black box of the Chinese policy process and showed how the reform process gradually deviated from its initial aims. The basis of the analysis was a conceptual framework that combined the structural and the process dimensions of central–local relations, and which encompassed both formal and informal institutions. A process tracing approach was employed to trace the relationship between institutions and agency in the sequential evolution of the reform. While the conclusions at the end of the chapters summarized the findings for each stage in empirical detail, I will now climb somewhat higher on the 'ladder of abstraction' (G. Sartori)[540] to condense these results into a set of more general findings regarding the impacts of central–local relations and agency on the outcome of each stage in the policy process. By analytically linking the sequences, a model of the RTFR process will be drawn up. Based on these findings, the second section outlines the theoretical and methodological value-added of the conceptual framework and research approach applied in this study. The final section returns to the issues of the peasant burden and rural governance to provide an assessment of the status quo and inspire further research.

A model of the Rural Tax and Fee Reform process

Given the conflicting needs to solve a problem while at the same time attending to the legitimate interests of the actors that are at once involved and affected by institutional change, I expected the information that underlay these processes to be biased towards the interest of the actors who presented them. For this reason, the agenda-setting process was understood as a struggle for constructing a narrative which served as the justification for the subsequent reform stages and thereby marked the boundaries of policy formulation. Similarly, policy formulation shaped the corridor of policy implementation, which in turn affected outcomes. Based on the findings of this study, the following sections will show that each stage in the Rural

Tax and Fee Reform process indeed delimitated the policy options available in the subsequent stage. Because each stage piled its dysfunctionalities on top of those inherited from the previous ones, the potential to attain the reform's initial goals was gradually diminished, and policy failure resulted. Each of the subsequent sections will elaborate on the limitations inherited from the previous stage and present several hypotheses regarding the respective roles of institutional path dependence and independent agency in affecting the outcomes of the respective sequence. The hypotheses are identified by means of a letter-number combination and will be united in a graphic representation of the model thus developed at the end of this section (Figure 8.1).

Sequence 0. Central–local relations and the peasant burden

The key question that formed the background for the present analysis was if institutional deficiencies or the moral degradation of local cadres were primarily responsible for heavy exactions from China's peasants and the related problems of bureaucratic sprawl and heavy indebtedness at the local levels. The findings of this study contradict the centre's notion that greed and predatory behaviour lay at the roots of these problems and posits that these symptoms were primarily rooted in the institutional set-up of central–local relations. In particular, three subordinate deficiencies can be identified:

First, the well known problem of unfunded mandates at the local level can be conceptualized as a mismatch between the delegation of administrative tasks and the recentralization of budgetary resources after 1994 **(A1)**. Second, the combination of a hierarchical prescription of development parameters and the stimulation of competition tended to amplify the problem, since the localities were required to not only spend more than they could afford, but often also more than was necessary **(A2)**. As a consequence, this combination of administrative delegation, fiscal recentralization and competition under hierarchy was clearly biased against agriculture-dependent and revenue-poor localities. Since they were unable to tap communal enterprises for taxes and off-budget revenue, they had to resort to borrowing and irregular exactions to meet the prescribed standards. Third, formal centralism in the fiscal and the policy-making realms converged with considerable informal revenue-generating and policy-making autonomy. This coexistence of formal top-down centralization and informal bottom-up decentralization was functional where the centre's and the localities' aims converged, i.e. in the broad realm of economic development. Where they did not, i.e. in the case of local predation and corruption, the centre was rather powerless to selectively withdraw these informal capacities **(A3)**.

In summary, the peasant burden was primarily a function of inequality in both local government revenue and peasant income coupled with an institutional set-up that fostered unbridled competition for resources and career advancement and which came at the price of a considerable loss of central control over local administrations.

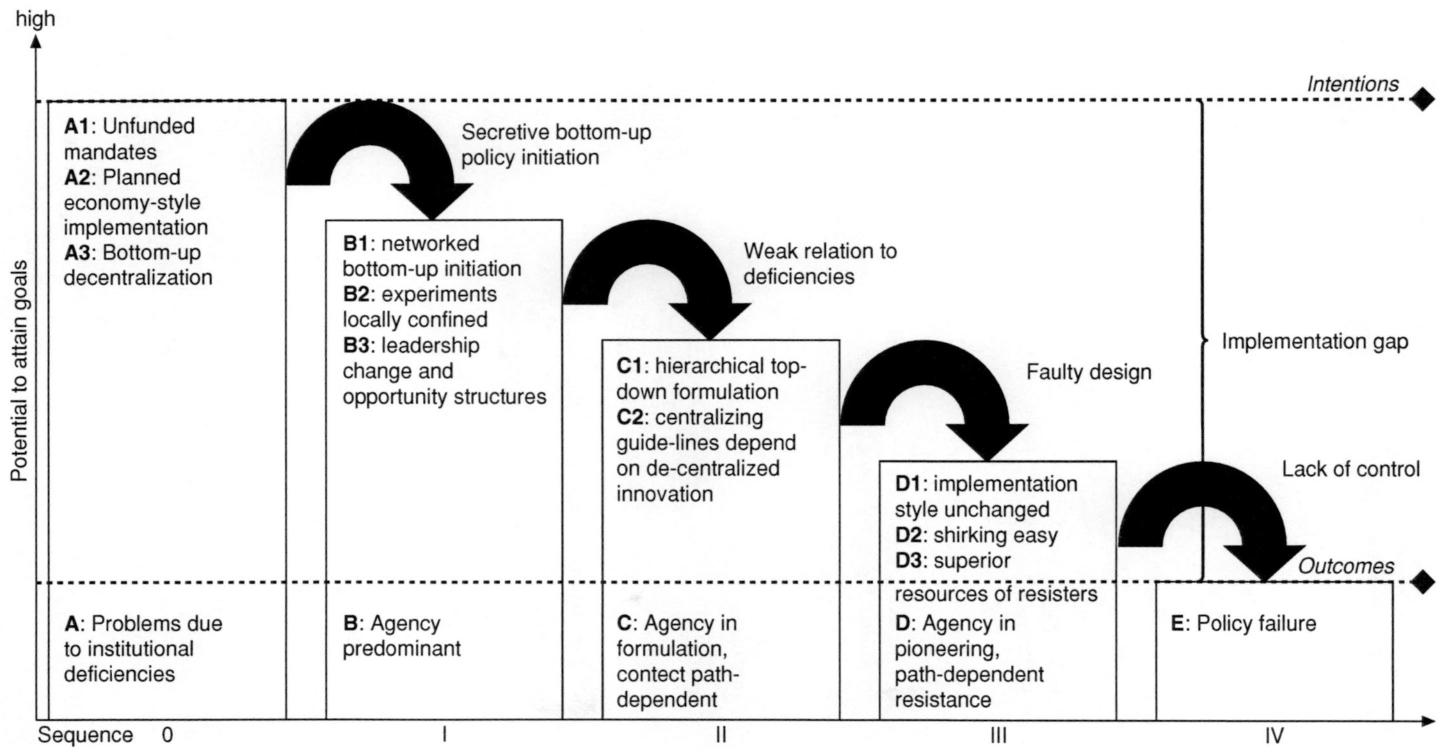

Figure 8.1 A model of the Rural Tax and Fee Reform process

Sequence I. Central–local relations and agenda-setting

Although a comprehensive and sustainable solution of the problems identified by the centre would have required equally comprehensive institutional changes, the agenda-setting process targeted only local-level institutions in parts of one province and did not address the more comprehensive deficiencies just discussed. Hence, the reform's potential to solve these problems was diminished. Why did this happen? The main reason, this study posits, is that the exclusive and secretive character of the agenda-setting process precluded a comprehensive discussion of the problem.

First, the agenda-setting process manifested itself as a decentralized experimentation process conducted by a very limited cast of actors bound in a cooperation-based network that spanned all administrative levels. It was tacitly approved and evaluated by the central government, but was withheld from the public agenda **(B1)**. Second, along with the limited stakeholder inclusion, the bottom-up character of this stage also explains why the dysfunctionalities laid out above were only partially addressed: local-level experimentation had to be confined to changing local structures **(B2)**. Third, the centre adopted the rural fiscal reform agenda not because the burden problem had become so salient, but as the result of a leadership change and because the new Premier desperately needed a quick success after several failures right at the beginning of his career **(B3)**. Hence, the reform at this stage was disposed towards short-term achievements and not long-term institutional change.

Sequence II. Central–local relations and policy formulation

The dynamics underlying the formulation of the RTFR policy removed the venture further away from its original intentions. Policy formulation was a hastened affair that again only included a small cast of actors, but one that was substantially different from the previous stage. In effect, a handful of central government cadres were given extraordinary competences to formulate an authoritative reform policy **(C1)**. On the other hand, the design that emerged from this process can be said to be almost path dependent in that it mirrored previous policies. It sought to recentralize central–local relations, but relied on decentralized policy innovation to achieve this aim. Hierarchical steering and the stimulation of competition were combined to enforce implementation **(C2)**. Like earlier policies, the reform was thus subject to the 'principality problem' of vague instructions coupled with the stimulation of competition. While this might have been a viable option had local agents been given incentives to do so, it became problematic when such incentives were absent. Normatively speaking, such a constellation requires that the local agents are given a role in policy formulation, and it presupposes a basic level of trust. However, suspicion prevailed, and the interpretative sovereignty lay with the centre, so this set-up became a tool of coercion. It was up to the centre to judge if the 'dispersed policy-making' did match up to the 'unified policy' or not. In other words: where the reform yielded good results, central policy was followed, and where it did not, local implementers could be easily blamed for deviation.

Sequence III. Central–local relations and policy implementation

The resulting design was therefore neither oriented towards comprehensively tackling the institutional sources of the problems they claimed to solve, nor did they integrate the viewpoints of the main stakeholders of the process, the county- and township-level governments. In addition, the reform programme was seriously flawed in design. First, as expected the combination of hierarchical steering and the stimulation of competition had detrimental effects on the finances especially of poor township- and county-level governments, who in turn were forced to continue irregular exactions and to incur further liabilities. Second, the complementary institutional restructuring policies were formulated so vaguely that they had to be operationalized by means of their own pilot reforms. As a result, they did not support, but were completely out of sync with the Main Reform. Third, the policy failed to address major historical problems such as the complicated local debt relations. Fourth, it was implicitly biased against small households with large landholdings. Fifth, it contained several loopholes and was subject to easy manipulation.

For these reasons, the resulting reform design was not very attractive to local implementers, and the centre was unable to enforce its proper implementation even though the implementation stage consisted of several sub-cycles in which the regulations were implemented, evaluated, reformulated or even changed. Although some pioneers could be motivated to contribute to the fine-tuning of the reform, the centre was unable to contain the habitual strategies of resistance. There are three main reasons for this.

First, reminiscent of previous policies, the process was inspired by a mix of hierarchical steering and the stimulation of competition among the implementing units in order to arrive at best solutions for the fine-tuning of the reform **(D1)**. The nationalizing of successful local experiments with individual Complementary Reform policies thus became an important component of policy fixing. However, as had been the case in the agenda-setting stage, these experiments could only address local institutional deficiencies and therefore did not do much to make the policy more attractive to the resisters. Implementation of these and other regulations was enforced by means of time-tested coercive sanctions against policy violators which, as shown below, proved not very effective.

Second, shirking as one form of resistance was made easy by the fact that the reform was at the same time under- and overregulated **(D2)**. It was under-regulated because of the vagueness of the individual policies, but over-regulated because not one, but almost 150 documents formed the basis of institutional restructuring. On the one hand, grassroots leaders were faced with difficult on-the-spot choices of which policies to favour over others, but on the other hand overregulation enabled them to shirk unwelcome policies, procrastinate their implementation or simply roll them off to the next lower level.

Third, the centre was unable to contain the resisters both because of their sheer mass, their distance from the centre and the superior resources they commanded **(D3)**. By withholding social services, they had a crucial leverage over rural stability which the protagonists in the centre did not want to see endangered.

Sequence IV. The outcome

In sum, the institutional deficiencies of the previous sequences accumulated to remove the outcomes of the reform far from its initial intentions. This section adds substance to the hypothesis that the initial policy has failed. Specifically, the impact of the Main Reform on aggregate peasant income had been negligible, but it had deprived the local- and grassroots-level administrations of vital resources needed to guarantee the continued provision of social services. The resulting situation was one where poor peasants enjoyed a reduction of their fiscal burdens, but still had to submit a substantially larger part of their net income to government coffers than rich peasants. In addition, grassroots-level administrations tended to manipulate the tax parameters or continued to levy illegal fees. Finally, many peasants were owed money by their townships and villages. When coercion against refusal to pay taxes was explicitly and effectively outlawed in 2002, they felt justified to resist exactions, an example which quickly spread. For the same reason, peasants could not be motivated to contribute funds to the 'one undertaking, one decision' programmes. This led to an even tighter revenue squeeze at the local level. As a consequence, counties intercepted transfer payments intended for the villages, and townships pocketed the surcharge to the agricultural tax that had replaced the village retention. Eventually, the villages were starved of revenue, county- and township-level debts had increased steeply, and local governments had become unable to meet their tax quotas. In sum, the agricultural tax had become dysfunctional as an instrument of revenue generation. As opposed to the official justification of this step presented by Premier Wen Jiabao, namely that the abolition of the agricultural tax as the second stage of a hitherto successful RTFR, this thesis has provided much evidence that the abolition was a necessary reaction to the failure of the original reform design.

Compared to previous attempts in history to rein in exactions from the peasants, the RTFR differed in two main aspects. First, as just discussed, the fiscal system was revamped to make exactions more transparent. Second, it entailed policies to improve and make more effective local governance. Territorial units were to be reduced, administrative structures simplified, superfluous government employees let go, transparency increased, business and politics separated, and fiscal capacities readjusted. The biggest successes were achieved in the merger of villages, townships and rural primary schools. In addition, a de-facto three-tiered administrative system consisting of the centre, the province and the county is in the making. The simplification of local administrative structures and the reduction of government employees were attempted, but due to local-level shirking only partially realized. Not only the centre, but also the regional and local governments utilized competition under hierarchy. The exceptions, of course, were township-level governments who are the lowest level in the Chinese administrative hierarchy and therefore cannot pass responsibilities further down. This explains why the counties transformed from reform objects to reform subjects. The provinces and cities had been expected to play a more important role in the RTFR, but they chose to impose the responsibility to implement the reform on the next lower level. As a consequence, the counties eventually became the primary locus where the Complementary Policies were specified, and the townships where they had to be implemented. Even where reform zeal was present, county governments did not seek to improve or alter their own organizational and employment structures, and neither did such pressure emanate from the provinces and cities. The result was that counties were clearly strengthened at the cost of severely weakened grassroots administrations. Even at these levels, personnel reduction only applied to over-the-payroll (*chaobian*) employees. Surplus cadres were not and could not be laid off. They were either sent into early retirement or 'redirected' (*fenliu*) to other units such as village schools, village Party branches or enterprises specifically established for absorbing surplus employees. A final way was to keep them off the job, but on the payroll. This was designed as an interim arrangement, but the odds are that these cadres will be provided for until their retirement. This explains the observations made in Chapter 7: The targeted reduction in personnel expenses could not be achieved, and administrative outlays kept increasing. Finally, apart from very isolated cases, transparency and supervision were not increased, expenditure structures not readjusted, business not separated from politics, and farmers' participation not furthered.

Clearly, the harder the targets of the respective policies were, the more likely was their implementation. The reduction of spatial units could be easily supervised by the centre. Personnel reductions and administrative simplification could also be quantified, whereas this was much more difficult with soft and sensitive targets such as transparency and participation. Coupled with a lack of incentives, a lack of resources and a lack of knowledge of how to implement such policies, this absence of clear benchmarks made soft targeted polices a non-priority at the local level.

Figure 8.1 visualizes the model that the previous sections have outlined. The grey columns recapitulate the main findings regarding the impacts of independent agency and the institutional framework of central–local relations on each sequence. Their diminishing size symbolizes how the reform's potential to attain its initial aims became gradually lessened by the deficiencies inherited from the previous stage. The main deficiency is spelled out and juxtaposed to the thick black arrow.

To sum up, it is posited that the exclusive and secretive bottom-up agenda-setting process addressed only selective local-level deficiencies and therefore was unable to come to terms with the complex and multi-faceted problems in the institutional design of central–local relations. Policy formulation was equally exclusive, but included none of the local stakeholders. A faulty policy design resulted which most local stakeholders were reluctant to implement. Finally, the centre was unable to enforce the proper implementation of its reform programme. This led to a substantive implementation gap, which is defined as the difference between intentions and outcomes.

In sum, then, the initial assumption that the failure of the RTFR was mainly the result of deficiencies in the structure and processes underlying central–local relations could not be upheld. The emerging picture is rather more complex. Agency has been instrumental in setting in motion processes at all stages of the policy cycle, but institutional factors were the main reason why the reform did not work out in the end. As demonstrated, however, this does not mean that institutional inertia or resilience can be made responsible for this outcome. Rather, the problem was that actors did not even attempt to address the deficiencies that had underlain central–local relations and which were reproduced in the design of the RTFR. It could now be argued that these rules could not be changed exactly because of their historical and social embeddedness. However, the analysis has shown that this was not the case. Rather, the protagonists at the local level merely aimed at changing local institutions, which the experiments achieved with some success in their respective contexts. The central protagonists sought to repeat this experience all over China. Even if this nationalization of the Main Reform did not achieve the desired results, it must be stressed that it set in motion a process that thoroughly changed the nature of the Chinese polity. This proves that institutions were not immune to change. As for the more abstract Complementary Reform policies, the centre merely set abstract guidelines that portrayed the polity it envisioned, but did not formulate concrete policies to further this aim. Ironically, the task of recentralization was left to the localities, who had only limited incentives and capacities to achieve this aim. Thus, dysfunctional institutions in the present case persisted not because of their social embeddedness, but because they were not appropriately addressed.

How did the central government evaluate the reform? As mentioned in the previous chapter, an explicit official evaluation does not exist, as the reform is portrayed as still ongoing. With respect to the RTFR, Premier Wen Jiabao stressed in early 2005 that the abolition of the agricultural tax was the second stage of the RTFR, which now had to be followed by a period of local-level consolidation.

In 2006, however, the 'Rural Tax and Fee Reform' was seamlessly transformed into the 'Rural Comprehensive Reform' (*nongcun zonghe gaige*), and the organizations affiliated with it also underwent a change of names. Nevertheless, inductions can be made regarding the centre's assessment of the reform. Specifically, policy documents summed up the achievements of each previous stage of the sub-cycles outlined in the previous sections and pointed out the problems that still had be solved. These implicit evaluations which, as this study has shown, highlighted the persistence of severe problems throughout all of these stages, stand in crass contrast to the generally positive views espoused by leading central politicians at public conferences, before the NPC and in the media. Thus, the presenting of public official evaluations bears resemblance to the dynamics observed in the agenda-setting process: public statements tend to be manipulative and not very representative of the discourses going on within the government.

Methodological reflection

The previous sections have formulated a model to explain why the reform outcomes did not match the initial intentions, and why the initial reform design eventually had to be abandoned. As a whole, the study has provided deep insights into the multi-faceted nature of central–local relations in China and the dynamics of a major structural reform attempt. This was made possible by means of two conceptual innovations and the combination of a sequential analysis with a fine-grained process tracing approach.

First of all, the study has shown that the impacts of the structure of intergovernmental relations on the quality of governance can be better understood if the process dimension of multi-level political steering is included into the analysis. The development of an enhanced typology of central–local relations is outside the scope of this single-case policy analysis, but would be definitely worth the effort. Comparative studies relating such a typology to the quality of grassroots governance would be equally fruitful. Second, the study followed Paul Hutchcroft's suggestion to examine not only on the formal parameters of the centralization–decentralization continuum, but to widen the focus to include informal institutions.[541] In this way, functional equivalents to formal decentralization could be identified and contrasted with conflicting laws and regulations in order to move closer to the roots of the observed systemic deficiencies.

As for the methodological approach, the combination of a sequence-based analysis and process tracing has proven profitable in two respects. First, it shed light on the intricate relationship between institutions and agency in the black box of the policy process and therefore precluded simplistic conclusions. Second, it illustrated how the range of options to attain certain goals gradually diminished as deficiencies and failures piled up throughout the reform stages. In this way, an analytic model has emerged that not only explained post-facto outcomes, but that also identified the critical junctures in the reform process and named the structural deficiencies they were especially prone to.

With regard to the Chinese context, this approach provides a useful complement to the 'experimental policy cycle' recently developed by Sebastian Heilmann.[542] While Heilmann has produced a very useful heuristic model to show 'how existing, and initially deficient, institutions can be put to work, transformed, or replaced for economic and social development in an open-ended process of institutional innovation that is based on locally generated solutions', the sequenced process-tracing approach can be applied to explain why experimental policy-making was abandoned or failed.

Outlook

Returning to the aims that had propelled the initial reform design, future research must now determine if the unintended abolition of rural taxes and levies along with the substantive recentralization of Chinese politics will work towards reducing the peasant burden in a sustainable way. This entails two related and rather normative questions. First, are the peasants now better off than before the reform? And second, has the readjustment contributed to improving rural governance in China?

In order to answer the first question, more micro-level research on the social and political consequences of rural income inequality after the reform must be conducted. Based on the observations made so far, I would posit that 'Huang Zongxi's law',[543] which underlay the dynamics of the dynastic cycle outlined in the introduction, has been broken. The Confucian scholar claimed that each tax reduction led to an increase in informal exactions that eventually exceeded the reduced amount. Given that China's peasants are now freed from all agricultural taxes and levies and are aware of their rights, it is unlikely that irregular levies can be re-imposed. The introduction of a land tax and the subjection of agricultural income to the income tax are unlikely as well. First of all, the governments' net receipts would be too low, given the high cost of fair, transparent and accountable tax collection. Second and relatedly, if the urban prerogative to a tax-exempt minimum income is extended to the rural areas, most peasants will have to be exempted from paying these taxes. More likely, the central government will continue its urbanization programmes to subject an increasing number of former rural inhabitants to urban taxation. This means that the peasant burden probably has been reduced in a sustainable way, but does that also mean that they are better off? As John Kennedy has shown, the tax and levy exemptions have directly and negatively affected the rural governments' provision of social services. This does not mean, however, that more income has simply been traded for fewer services. Again, rural income differences are a major intervening variable. While reduced services affect all rural inhabitants alike, the tax and fee exemptions have significantly benefited poorer peasants. In addition, prices for fertilizer, electricity and other factors which are essential for production and consumption have steeply risen, by far outpacing the increase in peasant income. Before this background, the question if the peasants as a whole are better off after the reform remains elusive and definitely merits further research.

In a related manner, the answering of the second question must go beyond merely asking if recentralization has improved governance. On the one hand, the basic tenets of the centralization versus decentralization discussion apply. Recentralization might help to lessen local state capture, clientelism and even corruption, but centralistic policy-making and – implementation comes at the price of responsiveness, especially in a polity as large, diverse and fragmented as China. On the other hand, it has been shown that although changes have been made with regard to the structure of the Chinese polity, changes in implementation styles have remained elusive. The basic and related problems of unfunded mandates and competition under hierarchy remain salient, and it is difficult to see how the increased fiscal transfers can effectively address the development problems in rural China under these circumstances. It seems likely that a large part of these transfers will continue to be wasted in projects that might satisfy superiors, but are not attuned to the necessities of the particular location. This insight might lie behind the central government's rationale of centralizing important rural social services by funding the extension to the rural areas of medical insurance and lowest cost-of-living payments. While this might contribute to solving the rural inequality problem espoused above, it also provides incentives for embezzlement and abuse. In sum, neither centralization nor decentralization can prevent local-level shirking. As the roles of the pioneers and resisters in this reform have shown, China's rural problems can only be solved if the local agents are motivated to subscribe to the centre's aims. They are more likely to do so if they can take part in policy formulation. Greater stakeholder inclusion not only in the agenda-setting, but also in the formulation of the policy would surely contribute to this aim.

Appendix Interviewees

Date of interview	*Interviewee's name* and position*	*Location**
2004		
July 23	Li Fan, head, China and the World think tank	Beijing
26	Professor, Central Party School	Beijing
28	Dang Guoying, researcher, CASS	Beijing
29	Shi Weimin, researcher, CASS	Beijing
August 12	Professor, Department of Law, Anhui University	Anhui Province, Hefei City
13	Former Party Secretary	Anhui Province, M Town
14	Head, Villagers' Committee	Anhui Province, Xiaogang Village
14	Two peasants	Anhui Province, Xiaogang Village
14	Party Secretary	Anhui Province, M Town
16	Li Kunsen, Vice-chairman, Anhui Province People's Congress	Anhui Province, Hefei City
17	Zhu Weimin, Bureau chief, Anhui Province RTFR LSG	Anhui Province, Hefei City
18	Zhou Weihan, head of Fiscal Department	Anhui Province, He County
19	Village Party Secretary	Anhui Province, He County, Yumin Village
September 6	Head of Organization Department	Shandong Province, L City
7	Party Secretary	Shandong Province, S Township
7	Dean of village primary school	Shandong Province, Z Village
8	Head of county government	Shandong Province, Y County
8	Head of county People's Congress	Shandong Province, Y County
9	Head of Legal Affairs Department	Shandong Province, Y County
9	Party Secretary	Shandong Province, D Township
10	Head of Villagers' Committee	Shandong Province, H Village
10	Peasant	Shandong Province, H Village
20	County Party Secretary	Hebei Province, W County
20	Head of county government	Hebei Province, W County
20	Head of Fiscal Department	Hebei Province, W County
20	Head of Legal Affairs Department	Hebei Province, W County
21	Village Party Secretary	Hebei Province, W Village
21	Peasant	Hebei Province, W Village
21	Town Party Secretary	Hebei Province, M Town
22	Village Party Secretary	Hebei Province, D Village
22	Township Party Secretary	Hebei Province, S Township
22	Deputy to the NPC	Hebei Province, S Township
23	Two peasants	Hebei Province, T Village

Date of interview	*Interviewee's name* and position*	*Location**
23	Three peasants	Hebei Province, T Village
24	Former head of village grain station	Hebei Province, T Village
24	Two peasants	Hebei Province, T Village
24	Two peasants	Hebei Province, T Village
25	Party Branch Secretary	Hebei Province, T Village
25	Head of Villagers' Committee	Hebei Province, T Village
27	Five migrant workers from Sichuan Province	Hebei Province, T Village
28	Former county vice-head	Hebei Province, P County
29	Entrepreneur	Hebei Province, P County
29	Five peasants	Hebei Province, X Village
29	Village doctor	Hebei Province, X Village
30	Former restaurant owner	Hebei Province, X Village
30	Head of Villagers' Committee	Hebei Province, X Village
October 2	Nine peasants	Hebei Province, X Village
3	Two peasants	Hebei Province, X Village
10	Vice-head of a Henan Province county government	Beijing
15	Journalist, *People's Daily*	Beijing
15	Journalist, Xinhua News Agency	Beijing
16	Migrant worker from Beijing suburb	Beijing
2006		
February 16	Journalist, *Gaige* magazine	Beijing
17	Migrant worker from Hubei Province	Beijing
20	Migrant worker from Anhui Province	Beijing
21	Zhao Shukai, researcher, DRF	Beijing
22	Migrant worker from Anhui Province	Yashou market, Beijing
24	Former member of Burden Reduction Department, Ministry of Agriculture, Anhui Province	Anhui Province, Hefei City
24	Zhang Deyuan, Professor, Anhui University	Anhui Province, Hefei City
24	Section chief of Anhui Bureau of Statistics	Anhui Province, Hefei City
24	Researcher, Anhui Academy of Science	Anhui Province, Hefei City
24	Zhu Shiqun, professor, Anhui University	Anhui Province, Hefei City
25	Vice-Party Secretary	Anhui Province, S Town
25	Vice-head of township government	Anhui Province, S Town
25	Head of Villagers' Committee	Anhui Province, X Village
27	Section chief of Anhui Bureau of Statistics	Anhui Province, Hefei City
27	Researcher, State Council Rural Development Research Centre	Anhui Province, Hefei City
March 1	Migrant worker from Anhui Province	Shanghai, Xiangyang Market
1	Migrant worker from Anhui Province	Shanghai, Xiangyang Market
4	Two migrant workers from Anhui Province	Shanghai, Xiangyang Market
4	Five migrant workers from Anhui Province	Shanghai, Xiangyang Market
5	Two migrant workers from Anhui Province	Shanghai
5	Two migrant workers from Anhui Province	Shanghai
6	Member of state Tax Department	Anhui Province, Bengbu City
6	Member of local Tax Department	Anhui Province, Bengbu City
7	Member of local Tax Department	Anhui Province, Bengbu City

Date of interview	*Interviewee's name* and position*	*Location**
8	Wu Li, professor, University of Economics	Anhui Province, Bengbu City
8	Head of county government	Anhui Province, W County
8	Former head of RTFR group	Anhui Province, W County
11	Vice-head of county government	Anhui Province, G County
11	Two heads of Villagers' Committees	Anhui Province, G County
11	Township Party Secretary	Anhui Province, G County
11	Township Party Secretary	Anhui Province, G County
12	Vice-head of county government	Anhui Province, G County
12	Head of Fiscal Department	Anhui Province, G County
14	Two peasants	Anhui Province, W Village
14	Dean of village school	Anhui Province, W Village
14	Five peasants	Anhui Province, W Village
15	Five peasants	Anhui Province, W Village
15	Peasant	Anhui Province, W Village
16	Party Secretary	Anhui Province, M Town
16	Head of township government	Anhui Province, M Town
16	Head of Organization Department	Anhui Province, M Town
16	Head of Fiscal Department	Anhui Province, M Town
16	Party Branch Secretary	Anhui Province, W Village
16	Three peasants	Anhui Province, W Village
17	Four peasants	Anhui Province, W Village
28	Zhu Gang, researcher, CASS	Beijing
28	Dang Guoying, researcher, CASS	Beijing
29	Journalist, *Gaige* magazine	Beijing
30	Gao Xinjun, researcher, Institute of Comparative Economics and Politics	Beijing
30	Lai Hairong, researcher, Institute of Comparative Economics and Politics	Beijing

*In order to protect my sources, most of them need to remain anonymous. In such cases, neither the name nor the interview location is specified.

Notes

1 The peasant burden and the dynamics of rural reform and development

1 D. Usher, 'The dynastic cycle and the stationary state', *American Economic Review* 79 (5), 1989, pp. 1031–44; P. Zhang, *Anhui nongcun shuifei gaige shijian yu tansuo* [Practice and exploration of Anhui's Rural Tax and Fee Reform], Beijing: Dangdai Chubanshe, 2001, pp. 36–7.

2 C., Riskin, R. Zhao and S. Li *China's Retreat from Equality: income distribution and economic transition*, Armonk NY: M. E. Sharpe, 2001, p. 75; World Bank, *World Development Indicators Database.* Online. Available HTTP: http://devdata.worldbank.org/data-query/ (accessed on December 1, 2007). The Gini index measures inequality, with 0 signalling the lowest and 1 the highest level of inequality. As Chinese income statistics tend to be 'polished', the gaps in both overall urban and rural wealth distribution are probably even more serious than these figures suggest. See S. Zhao, 'Cunzhuang yu fudan-yi xie difang nongmin fudan guozhong zhuangkuang de fenxi' [Villages and burden: an analysis of several locations where peasant burden is too heavy], *Guowuyuan fazhan yanjiu zhongxin xinxi wang*, October 31, 1999. Online. Available HTTP: http://www.drcnet.com.cn/DRCnet.Channel.Web/expert/showdoc.asp?doc_id = 137697 (accessed on October 20, 2009).

3 I am grateful to Claude Aubert for pointing out to me the reason for the decreasing yields.

4 *China Statistical Yearbook 2004*, Beijing: Zhongguo tongji chubanshe, 2004.

5 'China maintains food supply despite falling grain output', *Xinhua Online*, September 23, 2003. Online. Available HTTP: http://www2.chinadaily.com.cn/en/doc/2003–9/23/content_266735.htm (accessed on October 16, 2009). In 2004 and 2005, however, grain production had picked up again.

6 Z. Qu, X. Li and X. Wang, *Zhongguo shehui zhuyi xin nongcun jianshe yanjiu* [New rural construction in socialist China], Beijing: Shehui kexue wenxian chubanshe, 2006.

7 'CPC publishes key policy changes in party constitution', *People's Daily Online*, October 26, 2007. Online. Available HTTP: http://english.people.com.cn/90001/90776/90785/6291002.html (accessed on October 19, 2009).

8 W. Lam, 'Hu's New Deal and the Third Plenary Session of CCP's Seventeenth Central Committee', *China Brief* 8 (20), 2008. Online. Available HTTP: http://www.jamestown.org/programs/chinabrief/single/?tx_ttnews[tt_news] = 5230&tx_ttnews[backPid] = 168&no_cache = 1 (accessed on October 20, 2009).

9 Ibid.

10 D. Zhang, 'Wo kan "nongcun di san ci geming"' [My viewpoint on the 'Third Revolution in the Rural Areas'], *Shehui kexue zhanxian* 2003 (2), 2003, pp. 167–70. The RTFR was, not incidentally, followed in early 2006 by the campaign to

establish a 'New Socialist Countryside' (*shehui zhuyi xin nongcun*). In fact, the New Socialist Countryside programme, which had been brought up as early as 1998 (see Chapter 4), quite probably was a direct reaction to the failure of the initial RTFR policies (see Chapter 6).

11 Throughout this study, the county and city levels will be referred to as 'local', and the township and village levels as 'grass roots'. This departs from the Chinese practice of distinguishing only between the 'central' and the 'local' level (province and below). Often, the township and village levels are referred to as 'rural' (*xiangcun*) or 'grass roots' (*jiceng*). The obvious reason for this practice is the size of the alleged 'local' level: most of China's provinces match or even exceed the size of European states.

12 Guobanfa [2002] No. 25, *Guowuyuan bangonting guanyu zuohao 2002 nian guangda nongcun shuifei gaige shidian gongzuo de tongzhi* [Notification by the general office of the State Council concerning the orderly expansion of the RTFR pilot areas in 2002].

13 The 'three rural problems' refer to peasant poverty (*nongmin wenti*), to the stagnation of rural (social) development especially when compared to the cities (*nongcun wenti*), and to the lack competitiveness of the agricultural sector (*nongye wenti*). The expression is attributed to Li Changping, a former county chief, who wrote a letter to then Premier Zhu Rongji to draw attention to these issues and which contained the phrase that 'the villages are really poor, the peasants' life is really hard, and the situation of agriculture is really precarious' (*noncun zhen qiong, nongmin zhen ku, nongye zhen weixian*) (see C. Li, *Wo xiang zongli shuo shihua* [I spoke the truth to the Premier], Beijing: Guangming ribao chubanshe, 2002, preface).

14 See Zhongfa [2000] No. 7, *Zhonggong zhongyang, guowuyuan guanyu jinxing nongcun shuifei gaige shidian gongzuo de tongzhi* [Notification by the central committee of the Communist Party of China and the State Council concerning the advancement of work in RTFR pilot areas]. The distinction between the 'Main Reform' (*zhuti gaige*) and 'Complementary Reforms' (*peitao gaige*) is maintained throughout all policy documents concerning the RTFR.

15 Zhongfa [2000] No. 7.

16 Ibid.

17 Anhui Province, RTFR LSG, *Anhui sheng nongcun shuifei gaige qianhou nongmin fudan qingkuang biao* [Situation of peasant burden before and after the Rural Tax and Fee Reforms in Anhui Province], internal reference, 2003.

18 S. Zhu, H. Zhang and H. Yan, 'Nongcun shuifei gaige shidian he xiangcun guanli tizhi gaige genzong yanjiu baogao' [Follow-up research report on the Rural Tax and Fee Reform trial areas and rural administrative system reforms], *Jingji Yanjiu Cankao* 40, 2003, pp. 2–24.

19 Own calculations, based on Anhui Province, RTFR LSG, *Anhui sheng nongcun shuifei gaige qianhou nongmin fudan qingkuang biao* and *Anhui Statistical Yearbook 2004*, Hefei: Anhui Tongji Chubanshe, 2004.

20 L. Brandt, H. Liu, S. Rozelle and L. Zhang, *China. Rural Public Finance: the township perspective*, Annex 8, World Bank Report *Fiscal reform and role of the township*, Washington DC: World Bank, 2006; B. Wang and Y. Zhao, 'Nongcun shuifei gaige dui zhongxi bu xiangzhen caili yingxiang de shijian yanjiu – jiyu 4 sheng 8 xian chouyang diaocha shuju de fenxi' [Empirical research on the fiscal effects of the RTFR on central and western Chinese townships and towns], *Guanli Shijie*, No. 11 (2006).

21 J. J. Kennedy, 'From the Tax-for-fee reform to the abolition of agricultural taxes: the impact on township governments in north-west China', *China Quarterly* 189, 2007, pp. 43–59.

22 A quantitative analysis of the reform outcome is found in Christian Göbel, 'Measuring and explaining the distributive effects of the Rural Tax and Fee Reform in Anhui Province, China', *China Perspectives* 2008/02, pp. 58–79.
23 Guofa [2004] No. 21. *Guowuyuan guanyu zuohao 2004 nian shenhua nongcun shuifei gaige shidian gongzuo de tongzhi* [Notice by the State Council regarding doing well the intensivization of RTFR pilot work in 2004].
24 Own calculation from *Anhui Statistical Yearbook*, various years.
25 Own calculation from *China Statistical Yearbook*, various years.
26 State Council general office, *Report on the Work of the Government 2005*. Online. Available HTTP: http://www.gov.cn/english/official/2005-7/29/content_18351.htm (accessed on June 23, 2009).
27 L. C. Li, 'Working for the peasants? Strategic interactions and unintended consequences in the Chinese rural tax reform', *China Journal* 57 (2007), p. 90.
28 See D. Nohlen, *Wahlrecht und Parteiensystem. Zur Theorie der Wahlsysteme*, Opladen: Leske & Budrich, 2003, pp. 418–26, with respect to the design of electoral systems.
29 A useful overview is given in W. Merkel, *Systemtransformation. Eine Einführung in die Theorie und Empirie der Transformationsforschung*, Opladen: Leske & Budrich, 1999, chapter 6.
30 J. Brehm and S. Gates, *Working, Shirking, and Sabotage: bureaucratic response to a democratic public*, Ann Arbor MI: University of Michigan Press, 1997.
31 R. Yep, 'Can "tax for fee" reform reduce rural tension in China? The process, progress and limitations', *China Quarterly* 177, 2004, p. 68. This view is also taken by Bernstein and Lü. See T. P. Bernstein and X. Lü, 'Taxation without representation: peasants, the central and the local states in reform China', *China Quarterly* 163, 2000, p. 752, and T. P. Bernstein and X. Lü, *Taxation without Representation in Rural China*, Cambridge and New York: Cambridge University Press, 2003.
32 X. Gao, 'Nongcun shuifei gaige – redian wenti leng sikao' [The Rural Tax and Fee Reforms: cool thoughts on a hot problem], in Z. He (ed.) *Nongcun shuifei gaige – redian wenti leng sikao*, Beijing: Zhongguo bianyi chubanshe, 2004, pp. 180–94.
33 S. Zhu, H. Zhang and H. Yan, 'Nongcun shuifei gaige shidian he xiangcun guanli tizhi gaige genzong yanjiu baogao' [Follow-up research report on the Rural Tax and Fee Reforms trial areas and rural administrative system reforms], *Jingji Yanjiu Cankao* 2003 (40), 2003, pp. 2–24; D. Zhang, 'Shuifei gaige hou de de "bishui xiaoying"' [The 'tax enforcement phenomenon' after the RTFR], in C. Li and L. Dong (eds), *Shuifei gaige beijing xia de xiangzhen tizhi yanjiu*, Wuhan: Hubei renmin chubanshe, 2004, pp. 326–30.
34 Zhu, Zhang and Yan, 'Nongcun shuifei gaige shidian he xiangcun guanli tizhi gaige genzong yanjiu baogao'.
35 L. C. Li, 'Provincial discretion and national power: investment policy in Guangdong and Shanghai, 1978–1993', *China Quarterly* 152, 1997, pp. 778–804.
36 Ibid.
37 L. C. Li, 'Differentiated actors: central and local actors in China's rural tax reforms', *Modern Asian Studies* 40 (1), 2006, pp. 173–4.
38 Ibid., p. 172.
39 Ibid. Many other examples might be mentioned here, such as the introduction of the Household Responsibility System (L. T. White, *Unstately Power*, Armonk NY: M. E. Sharpe, 1998), village elections (D. Kelliher, 'The Chinese debate over village self-government', *China Journal* 37, 1997, pp. 63–86), the privatization of the rural economy (T. Heberer, *Doing Business in Rural China: Liangshan's new ethnic entrepreneurs*, Seattle WA: University of Washington Press, 2007;

T. Heberer, *Private Entrepreneurs in China and Vietnam: social and political functioning of strategic groups*, Leiden and Boston MA: Brill, 2003; B. Alpermann, *Economic Transformation and State Capacity: the case of the Chinese cotton sector*, dissertation, University of Cologne, 2006) and the reform of state-owned enterprises (S. Heilmann, 'Policy experimentation in China's economic rise', *Studies in Comparative International Development* 43 (1), 2008, pp. 1–26).

40 Heilmann, 'Policy experimentation in China's economic rise', p. 22.

41 Ibid., p. 23.

42 J. Oi, *Rural China Takes Off: institutional foundations of economic reform*, Berkeley CA, etc.: University of California Press, 1999.

43 L. C. Li, 'Working for the peasants? Strategic interactions and unintended consequences in the Chinese rural tax reform', *China Journal* 57, 2007, p. 100.

44 See also Chapter 3.

45 X. Ma, H. Huang and C. Jiang, *Wo guo nongcun shuifei gaige yanjiu* [The Rural Tax and Fee Reform in China], Beijing: Zhongguo ji hua chu ban she, 2002.

46 Ibid., pp. 94–106.

47 P. K. Bardhan and D. Mookherjee, 'The rise of local governments: an overview', in P. K. Bardhan and D. Mookherjee (eds), *Decentralization and Local Governance in Developing Countries: a comparative perspective*, Cambridge MA: MIT Press, 2006, pp. 4–15.

48 H. Fuhr, 'Institutional change and new incentive structures for development: can decentralization and better local government help?', *Welttrends* 25 (4), 2000, pp. 21–51.

49 P. Bardhan, 'Decentralization of governance and development', *Journal of Economic Perspectives* 16 (4), 2002, p. 185.

50 P. Hutchcroft, 'Centralization and decentralization in administration and politics: assessing territorial dimensions of authority and power', *Governance: an International Journal of Policy and Administration* 14 (1), 2001, p. 23.

51 L. J. Diamond, *Developing Democracy: toward consolidation*, Baltimore MD: Johns Hopkins University Press, 1999, pp. 132–8 and 55–7; World Bank, *East Asia Decentralizes: making local government work*, Washington DC: World Bank, 2005, pp. 239–41.

52 Bernstein and Lü, *Taxation without Representation in Rural China*, p. 56. Chen and Chun provide more examples: G. Chen and T. Chun, *Zhongguo nongmin diaocha* [An inquiry into the situation of China's farmers], Beijing: Renmin wenxue chubanshi, 2004, chapter 7.

53 A. MacIntyre, M. H. Khan and K. S. Jomo, 'Funny money: fiscal policy, rent-seeking and economic performance in Indonesia', in M. H. Khan and K. S. Jomo (eds), *Rents, Rent-seeking and Economic Development: theory and evidence in Asia*, Cambridge: Cambridge University Press, 2000, p. 251.

54 Heilmann, 'Policy experimentation in China's economic rise', p. 2.

55 A still relevant overview of the most common typologies can be found in P. Kenis and V. Schneider, 'Verteilte Kontrolle. Institutionelle Steuerung in modernen Gesellschaften', in P. Kenis and V. Schneider (eds), *Organisation und Netzwerk. Institutionelle Steuerung in Wirtschaft und Politik*, Frankfurt and New York: Campus Verlag, 1996, p. 19.

56 Ibid., p. 20.

57 See C. Hood, *The Art of the State: culture, rhetoric, and public management*, Oxford and New York: Clarendon Press, 1998.

58 Ibid., pp. 24–5.

59 Brehm and Gates, *Working, Shirking, and Sabotage: bureaucratic response to a democratic public.*

60 Hood, *The Art of the State*, pp. 55–60.

61 Ibid., p. 28.
62 A notable exception is a heuristic model of public policy-making in China contributed by Sebastian Heilmann (which, however, does not apply the categories just outlined). His model, which he calls an 'experimentation-based policy cycle', illustrates how a cycle of experimental policy implementation and refinement can contribute to produce important policy innovations in a highly restrictive structural environment. See Heilmann, 'Policy experimentation in China's economic rise'.
63 A fascinating account of the historical role that experiments and emulation have played in Chinese policy-making is found in S. Heilmann, 'From local experiments to national policy: the origins of China's distinctive policy process', *China Journal* 59, 2008, pp. 1–30.
64 Li, 'Working for the peasants?', p. 93.

2 Bureaucratic sprawl, government debt and the peasant burden: how serious?

65 On the different workforces and the problems of terminology see K. E. Brodsgaard, 'Institutional reform and the *bianzhi* system in China', *China Quarterly* 170, 2002, pp. 361–86, and J. P. Burns, '"Downsizing" the Chinese state: government retrenchment in the 1990s', *China Quarterly* 175, 2003, pp. 775–802.
66 For a detailed description and typology of the term 'cadre' see T. Heberer, 'Das politische System der VR China im Prozess des Wandels', in C. Derichs and T. Heberer (eds), *Einführung in die politischen Systeme Ostasiens*, Opladen: Leske & Budrich, 2003, pp. 62–5, 19–121, and Y. Zhong, *Local Government and Politics in China: challenges from below*, Armonk NY: M. E. Sharpe, 2003, pp. 94–126.
67 See for example Z. Liu, 'Zhongguo zhengfu jigou cunzai de zhuyao wenti' [The main problems in China's government structure], *Zhanlüe yu guanli* 5 (36), 1999, pp. 90–5.
68 S. Zhao, 'Xiangzhen gaige: jiantao yu zhanwang' [Township and town reforms: exploration and prospects], in X. Wu, X. Lu and P. Li (eds), *Zhongguo shehui xingshi fenxi yu yuce*, Beijing: Shehui kexue wenxian chubanshe, 2005, p. 77.
69 In a way, this comparison does not make much sense, as the Chinese government's share in the country's GDP is much lower than that of the OECD countries, as is GDP per capita, which is even more relevant to the comparison. Accordingly, even the 'big' ratio of 3.6 per cent is very small by OECD standards. In 1998, two-thirds of all OECD countries surpassed the 6 per cent mark. With more than 13 per cent, Denmark has the highest ratio of public servants, Korea, with 2 per cent, the lowest. Simply put, however, given its large population, its low per capita income and the government's small share of the GDP, China cannot afford even that comparatively small ratio of public servants. Data calculated on the base of OECD, *Public Sector Pay and Employment Data: total public employment*, online, available HTTP: http://www.oecd.org/dataoecd/27/17/1827384.xls (accessed on October 2, 2009), and OECD, *Total Population for OECD countries*, online, available HTTP: http://www.oecd.org/dataoecd/62/38/35267227.pdf (accessed on October 2, 2009).
70 Burns, '"Downsizing" the Chinese state', pp. 787–90.
71 Ibid., p. 792.
72 See Q. Xie, *Xian zhengfu guanli: Wanning Xian diaocha* [County government management: an investigation of Wanning County], Beijing: Guangbo dianshi chubanshe, 1994, pp. 96–101.
73 Zhao, 'Xiangzhen gaige: jiantao yu zhanwang', p. 77.

74 G. Du, 'Jigou gaige chedi, bixu dangzheng fenkai' [For thorough organizational reform, Party and government must be separated], *Jiancha Ribao*, January 16, 2003. Online. Available HTTP: www.jcrb.com/zyw/n6/ca12185.htm (accessed on October 2, 2009).

75 For more details on the administrative restructuring in 1982 and 1983 see J. Yu, *Yuecun zhengzhi: Zhuanxingqi Zhongguo xiangcun zhengzhi jiegou de bianqian* [Politics in Yue Village: the changes in rural political structures in transforming China], Beijing: Shangwu yinshu guan, 2001, chapter 4, and X. Pu, *Zhonghua renmin gongheguo zhengzhi tixi* [The political system of the People's Republic of China], Shanghai: Shanghai renmin chubanshe, 1999, pp. 36–46.

76 On the change of property rights to local governments see M. Taube, *Institutionelle Lebenszyklen im chinesischen Transformationsprozess. Aufstieg und Fall der Township Village Enterprises*, Duisburg: Diskussionsbeiträge der Mercator School of Management der Universität Duisburg-Essen 322, 2007; C. Chang, B. P. McCall and Y. Wang, 'Incentive contracting versus ownership reforms: evidence from China's township and village enterprises', *Journal of Comparative Economics* 31 (3), 2003, pp. 414–28; S. H. Whiting, J. McMillan and B. Naughton, 'Contract incentives and market discipline in China's rural industrial enterprises', in J. McMillan and B. Naughton (eds), *Reforming Asian Socialism: the growth of market institutions*, Ann Arbor MI: University of Michigan Press, 1996, pp. 63–110; S. P. S. Ho, *Rural China in Transition: non-agricultural development in rural Jiangsu, 1978–1990*, Oxford and New York: Oxford University Press, 1994; J. C. Oi and A. G. Walder, *Property Rights and Economic Reform in China*, Stanford CA: Stanford University Press, 1999; J. C. Oi, *Rural China Takes Off: institutional foundations of economic reform*, Berkeley CA and London: University of California Press, 1999.

77 Z. Lan, 'Local government reform in the People's Republic of China: stipulations, impact, cases and assessment', *Chinese Public Administration Review* 1 (3–4), 2002.

78 Ibid., p. 213.

79 T. C. Lam and J. L. Perry, 'Service organizations in China: reform and its limits', in P. N.-S. Lee and C. W.-H. Lo (eds), *Service Organizations in China: reform and its limits*, Westport CT: Quorum Books, 2001, pp. 19–40.

80 C. Wong, C. J. Heady and W. T. Woo, *Fiscal Management and Economic Reform in the People's Republic of China*, Hong Kong and New York: Oxford University Press, 1995, p. 121.

81 Zhao, 'Xiangzhen gaige: jiantao yu zhanwang', p. 67.

82 X. Chen, *Zhongguo xian xiang caizheng yu nongmin zengshou wenti yanjiu* [Research on public finances in China's counties and townships and the problem of increasing peasant income], Taiyuan: Shanxi jingji chubanshe, 2003, p. 281.

83 Calculated from *China Statistical Yearbook 2005*, Beijing: Zhongguo tongji chubanshe, 2005.

84 Chen, *Zhongguo xian xiang caizheng yu nongmin zengshou wenti yanjiu*, p. 177.

85 Interview, Anhui Province, G County, March 14, 2006.

86 Source: M Town, Finance Office, *2006 nian M zhen caizheng shouzhi* [*Budgetary revenues and expenditures of M Town for 2006*], internal reference, 2006.

87 This is the official figure from the Ministry of Agriculture for the year 2000. It is quoted in J. Oi and S. Zhao, 'Fiscal crisis in China's townships: causes and consequences', in E. J. Perry and M. Goldman (eds), *Grassroots Political Reform in Contemporary China*, Cambridge MA: Harvard University Press, 2007, pp. 75–96, p. 75.

88 Interview, Anhui Province, M Town, March 16, 2006.

89 Quoted in D. Ding, 'Wo guo xiang (zhen) zun fuzai wenti de chengyin ji qi huajie: Anhui Sheng Suzhou Shi xiang (zhen) cun fuzai qingkuang de diaocha', *Jingji Yanjiu Cankao* 72 (1744), 2003, p. 41.
90 Interview, Beijing, March 28, 2006.
91 G. Zhu, *Xiangcun zaiwu guimo yu zaiwu fengxian* [The extent and dangers of rural debt], Beijing: manuscript, 2006.
92 Interview, Beijing, March 28, 2006.
93 N. Lardy, 'China's worsening debts', *Financial Times*, June 22, 2001. Online. Available HTTP: http://www.brookings.edu/opinions/2001/0622china_lardy.aspx (accessed on November 2, 2007).
94 Y. Cai, 'Between state and peasant: local cadres and statistical reporting in rural China', *China Quarterly* 163, 2000, pp. 783–805.
95 Bernstein and Lü, *Taxation without Representation in rural China*, pp. 112–13.
96 Interview, Beijing, February 21, 2006.
97 *China Statistical Yearbook 2006*, Beijing: Zhongguo tongji chubanshe, 2006.
98 See Treaty on European Union, Protocol on the Excessive Deficit Procedure, Art. 1.
99 Lardy, 'China's worsening debts'.
100 CIA, *The CIA World Factbook on CD-ROM*, Minneapolis MN: Quanta Press, 2007. Only the Lebanon surpasses Japan.
101 H. Li, 'Finance in rural China', in Rural Development Institute of the Chinese Academy of Social Sciences (ed.), *Finance in Rural China*, Beijing: Foreign Languages Press, 2006, p. 98. See also 'Bail-out of debt-laden Agriculture Bank mulled over', *Xinhuanet*, May 26, 2006. Online. Available HTTP: http://news.xinhuanet.com/english/2006–05/26/content_4604905.htm (accessed on October 19, 2009).
102 'Agricultural Bank faces tough restructuring path', *Xinhua Online*, May 29, 2006. Online. Available HTTP: http://news.xinhuanet.com/english/2006–05/29/content_4615017.htm (accessed on October 19, 2009).
103 S. Tanigaki, 'Japanese Economy Revived: challenges and the way ahead: fiscal consolidation is the biggest remaining challenge', paper presented at Japan Society, New York, January 9, 2006.
104 Zhongguo Shekeyuan Caimaosuo Ketizu, *Difang zhengfu zhaiwu: laiyuan, goucheng, yingxiang ji huajie duice* [*Local government debt: origins, composition, influence and clearance counterpolicies*], Beijing: CASS.
105 Ibid.
106 Ministry of Finance 1999, quoted in X. Chen, J. Han and Y. Zhao, *Zhongguo nong cun gong gong cai zheng zhi du: li lun, zheng ce, shi zheng yan jiu* [Public finance in rural China: theory, policies, empirical research], Beijing Shi: Zhongguo fazhan chubanshe, 2005, p. 76.
107 The comparison is made by simply multiplying the number of Anhui's townships and towns with the average township-level debt figure provided for each prefecture. The absolute township-level debt figures thus obtained are added up to yield the provincial total of 6.17 billion yuan. The same procedure is conducted with village-level data, which add up to 3.4 billion yuan.
108 In 2003 the combined revenue of Anhui's fifty-six counties and five county-level cities added up to 7.2 billion yuan RMB. Calculated from *Anhui Statistical Yearbook 2004*, Hefei: Anhui Tongji Chubanshe, 2004.
109 Zhu, *Xiangcun zaiwu guimo yu zaiwu fengxian*, p. 7.
110 Ibid., pp. 9–10.
111 T. Zhou, *Zhongguo zhengzhi tizhi gaige* [Political system reform in China], Beijing: Zhongguo shuili shuidian chubanshe, 2004.
112 Ibid.

113 Author's interview, Anhui Province, M Town, March 16, 2006.
114 Zhu, *Xiangcun zaiwu guimo yu zaiwu fengxian.*
115 Ibid.
116 see ibid., pp. 16–22.
117 Analysis based on data derived from ibid., pp. 9–10 and 11.
118 As correlation analysis reveals only the degree in which the variations in two data sets correspond to each other, not the direction of causality, an alternative hypothesis deriving from these data would be that peasant burden determines local debt. However, nothing seems to support this hypothesis.
119 Author's interview, Beijing, February 21, 2006. Some of these points are also found in an English-language article co-authored with Jean Oi: J. Oi and S. Zhao, 'Fiscal crisis in China's townships: causes and consequences', in E. J. Perry and M. Goldman (eds), *Grassroots Political Reform in Contemporary China*, Cambridge MA: Harvard University Press, 2007.
120 This section is based on figures provided in X. Zhou, *'Xiang xai xian guan' gaige de shehuixue fenxi: yi Anhui Sheng G Xian wei li* [A sociological analysis of the 'county administers township finances' reform: a case study of Anhui's G County], unpublished Master's thesis, Anhui University, 2005, p. 15.
121 Author's interview, Anhui Province, G County, March 11, 2006. The 'tax division system' and the 'two columns' of education will be explained in the next chapter.
122 Bernstein and Lü, *Taxation without Representation in rural China*, p. 66.
123 See Zhao, 'Cunzhuang yu fudan-yi xie difang nongmin fudan guozhong zhuangkuang de fenxi'; Cai, 'Between state and peasant: local cadres and statistical reporting in rural China'; Bernstein and Lü, *Taxation without Representation in rural China*, p. 113.
124 Interview, Beijing, March 28, 2006. See also the section on local debt above. A hypothesis which stands to be tested is that excessive burden, as it is defined below, is indeed less of a problem in eastern China. As media reports indicate, the illegal reacquisition of agricultural land for economic purposes seems to have been the main issue of peasant discontent there.
125 Anhui, for example, had increased the 1963 'average harvest' base of 7.8 billion kg to 15.8 billion kg in 1977, but at the same time reduced the tax rate from 15.5 per cent to 7 per cent. Since then, neither had been readjusted. Thus, in 1994, the real tax rate paid by the farmers in 1994 was only 3.09 per cent for Anhui and 2.5 per cent for all of China. K. He and L. Sun (eds), *Zhongguo nongcun shuifei gaige chutan* [A first exploration of the Rural Tax and Fee Reform in China], Beijing: Zhongguo Zhigong Chubanshe, 2000. The next chapter provides further explanations on the agricultural taxes.
126 The village retention financed three items, the township comprehensive levy five (see next chapter). Thus they became commonly known as *santi wutong*, 'three retentions and five comprehensives'.
127 Unless otherwise indicated, I follow the practice of the national and provincial statistical yearbooks and refer to the 'four agricultural taxes' (*nongye sishui*, i.e. agricultural tax, tax on agricultural special products, slaughter tax and animal husbandry tax) simply as 'agricultural taxes'. Whereas the yearbooks speak of 'agricultural tax and related revenue', for stylistic reasons I prefer the aforementioned term. The difference between the agricultural tax and the tax on agricultural special products is that the latter was higher and levied on cash crops, whereas the former was lower and applied to all other produce. In reality, however, either tax was applied quite indiscriminately, and sometimes both taxes were raised on the same product. In fact, only in 2000 did a regulation explicitly specify which goods belonged to the 'special products' category. See Caishuizi [2000] No. 67, *Caizhengbu, Guojia shuiwu zongju guanyu nongcun shuifeigaige shidian diqu nongye techanshui zhengce de tongzhi* [Notification by

the Ministry of Finance and the National Tax Administration Office concerning the policies towards the special products tax in Rural Tax and Fee Reform trial areas]). As the name indicates, the slaughter tax had to be paid on each slaughtered animal and was set at 10 per cent ad valorem of the weight of the slaughtered hog, cattle or sheep. However, the slaughter tax was generally apportioned and levied per capita, which caused had caused great resentment among the peasantry (see Chapter 3). Hence it was the first tax to be completely rescinded in the RTFR pilot areas in May 2000. See Caishuizi [2000] No. 71, *Caizheng bu, guojia shuiwu zongju guanyu nongcun shuifei gaige shidian diqu quxiao tuzaishui de tongzhi* [Notification by the Ministry of Finance and the State Tax General Department regarding the abolition of the slaughter tax in the rural tax and Fee Reform pilot regions]. Finally, the animal husbandry tax was mainly levied in regions where animal husbandry prevailed over arable, as in the following provinces and autonomous regions: Inner Mongolia, Sichuan, Shaanxi, Gansu, Qinghai, Ningxia and Xinjiang. The tax rate was fixed by the provincial governments. In 2005 it had been rescinded in all of China. A good overview and explanation of China's local taxes is to be found in Z. Liu, *Zhongguo difang shuizhi gaijian* [Local taxation in China], Beijing: Zhongguo jingji chubanshe, 2005.

128 F. Guo, *Nongcun shuifu yu nongmin fudan* [The Agricultrural Tax and peasant burden], Beijing: Jingji ribao chubanshe, 2003, p. 40.

129 Cited in Chen, Han and Zhao, *Zhongguo nongcun gonggong caizheng zhidu*, p. 99. Figures for 'societal burden' and corvée labour are not given. Instead, 'other fees' functions as an unexplained residual category.

130 T. Xiao and C. Li, *Zhongguo xiangcun baogao: zhengfu xingwei yu xiangcun jianshe yanjiu* [A report on rural China: researching government behaviour and rural construction], Shanghai: Xu lin chubanshe, 2005, pp. 19–20.

131 The *China Statistical Yearbook* (used by the mentioned contributor as well) states average peasant income for 1993 as 921.60 yuan RMB, that for 1998 at 2,162 yuan RMB. Accordingly, the level of the *santi wutong* relative to average peasant income decreased from 4.84 per cent in 1993 to 3.89 per cent in 1998. See *China Statistical Yearbook 1994*, Beijing: Zhongguo tongji chubanshe, 1994, *China Statistical Yearbook 1999* and Xiao and Li, *Zhongguo xiangcun baogao*, pp. 19–20.

132 Xiao and Li, *Zhongguo xiangcun baogao*, p. 41, and *China Statistical Yearbook*, various years.

133 Chen, *Zhongguo xian xiang caizheng yu nongmin zengshou wenti yanjiu*, p. 129.

134 Ibid., p. 117.

135 Zhu Gang's study confirms this. In the central Chinese cases, the ratio of average peasant burden to income varied between 0.47 and 43.91 per cent, in the western Chinese cases from 0.08 to 76.46 per cent. Zhu, *Xiangcun zaiwu guimo yu zaiwu fengxian*, p. 11.

136 Chen, *Zhongguo xian xiang caizheng yu nongmin zengshou wenti yanjiu*, p. 118.

137 In the last two weeks of 2004, the author spent one week each in two villages belonging to Hebei's Pingxiang County. In the last two weeks of March 2006, he spent one week each in two villages belonging to central Anhui's Guzhen and Wuhe Counties. In those weeks he conducted formal and informal interviews with peasants and village cadres.

138 Xiao and Li, *Zhongguo xiangcun baogao*, p. 331.

3 The institutional sources of the peasant burden

139 Ibid., emphasis in the original.

140 R. Work, 'Overview of Decentralisation Worldwide: a Stepping Stone to Improved Governance and Human Development', paper presented at second

International Conference on Decentralisation ('Federalism: The Future of Decentralizing States?'), Manila, Philippines, July 25–27, 2002.

141 B. Baldi, 'Beyond the federal–unitary dichotomy', *Institute of Governmental Studies Working Paper* 99 (7), 1999, p. 13.

142 Ibid.

143 Work, 'Overview of decentralisation worldwide'.

144 R. D. Ebel and S. Yilmaz, *Concept of Fiscal Decentralization and Worldwide Overview*, Washington DC: World Bank, 2002.

145 Baldi, 'Beyond the federal–unitary dichotomy', p. 13.

146 D. Treisman, 'Political decentralization and economic reform: a game-theoretic analysis', *American Journal of Political Science* 43 (2), 1999, p. 490.

147 See D. L. Shambaugh, *Modernizing China's Military: progress, problems, and prospects*, Berkeley CA: University of California Press, 2002, chapter 4.

148 Oi, *Rural China Takes Off.*

149 An overview of various quantitative indicators is given in S. Powell, *Agricultural Reform in China: from communes to commodity economy, 1978–1990*, Manchester and New York: Manchester University Press, 1992, p. 36 (table 5).

150 Ho, *Rural China in Transition*, gives a good overview over the development of TVEs in the 1980s.

151 W. A. Byrd and Q. Lin, *China's Rural Industry: structure, development, and reform*, Oxford and New York: Oxford University Press, 1990, provide a valuable empirical account of the first phase of TVE development.

152 J. Fewsmith, *China since Tiananmen: the politics of transition*, Cambridge and New York: Cambridge University Press, 2001, pp. 37–8.

153 Ibid., pp. 55–63.

154 Ho, *Rural China in Transition, 1978–1990*, pp. 2–3.

155 See *China Statistical Yearbook 1997.*

156 E. Friedman, P. Pickowicz and M. Selden, *Chinese Village, Socialist State*, New Haven CT: Yale University Press, 1991, provide an excellent account of these dynamics.

157 The resemblance of such target-setting activities to policy implementation during the Great Leap Forward is of course not lost on observers. The comparison is made by T. P. Bernstein and X. Lü, *Taxation without Representation in Contemporary Rural China*, Cambridge: Cambridge University Press, 2003, p. 95, as well as Chen Guidi and Wang Chuntao. Chen and Wang present several enlightening case studies that illustrate the problem well (Chen and Chun, *Zhongguo nongmin diaocha*, pp. 196–204). About half the book's chapters have been published as an English-language monograph: G. Chen and T. Chun, *Will the Boat sink the Water? The life of China's peasants*, New York: Public Affairs, 2006).

158 S. H. Whiting, *Power and Wealth in Rural China: the political economy of institutional change*, New York and Melbourne: Cambridge University Press, 2001, p. 110.

159 Quoted in ibid., p. 119.

160 Ibid., pp. 110–11.

161 Author's interview, Anhui Province, April 12, 2006. This problem is also confirmed by a survey which asked several rural credit cooperatives to assess the quality of their customers. (Unfortunately, the number of respondents is not supplied.) While the overwhelming majority saw no or only slight danger in extending micro- and agricultural credit and lending to peasants, more than 80 per cent found lending to township-level governments a 'rather great risk' (*fengxian jiao da*) or 'basically not possible' (*jiben shang bu neng fangkuan*). As for loans to TVEs, every single respondent answered in these two categories.

No one considered such loans a 'rather small risk' (*fengxian jiao xiao*) or no risk at all (G. He, *Zhongguo nongcun jinrong fazhan yu zhidu bianqian* [The development of rural credit in China and institutional change], Beijing: Zhongguo caizheng jingji chubanshe, 2005, p. 167).

162 See *China Statistical Yearbook*, various years.

163 The TVEs' rising credit demand is reflected also by a second indicator, the composition of debtors to the RCCs, whose primary purpose, it should be recalled, is the financing of agriculture. Up to 1982, when credits for agricultural projects formed more than 60 per cent of the portfolio, it served this purpose well. After that year, the percentage of such credits declined rapidly. In 1988, they had dropped to around 40 per cent, and in 1994 to just 19.4 per cent. The percentage of credits for TVEs, however, increased steeply, from 38.1 per cent in 1980 to around 55 per cent in 1994. That year, ominous 'other credits' had risen from less than 10 per cent to almost 30 per cent. He, *Zhongguo nongcun jinrong fazhan yu zhidu bianqian*, p. 129. From 1994 on, the situation seemingly improved, with the percentage of agricultural credits slowly increasing to around 35 per cent, and loans to TVEs decreasing to around 45 per cent in 2000. 'Other credits' formed the remaining 20 per cent. Ibid., p. 167. The main reason for this decline is not increased lending to support agriculture, but the formation of the Agricultural Development Bank of China (Zhongguo nongye fazhan yinhang, ADB) in 1994, which extended loans to TVEs engaged in the food sector.

164 Author's interview, Anhui Province, G County, March 8, 2006.

165 World Bank, *China: National Development and Sub-national Finance: a review of provincial expenditures*, World Bank Report 22951-CHA, 2002.

166 Ibid.

167 Ibid.

168 Bernstein and Lü, *Taxation without Representation in Rural China*, p. 56.

169 Ibid., p. 91.

170 Chen and Chun, *Zhongguo nongmin diaocha*, pp. 196–204. The authors provide more examples in the following pages.

171 Author's interview, Shandong Province, September 6, 2004. The headmaster claimed that he had paid the replica out of his own pocket.

172 Bernstein and Lü, *Taxation without Representation in Rural China*, p. 56.

173 Author's interview, Hebei Province, September 28, 2004.

174 Author's interview, Shandong Province, September 9, 2004. In the context of another marginalized group, China's urban poor, Dorothy Solinger even implies that the centre's interest lies more in retaining than in changing such discriminatory structures, which are primarily geared towards producing economic rent. Unrest is quelled not by institutional change, but by the application of 'weapons of the state', an amalgam of alliances, propaganda, forged statistics and coercion. D. Solinger, 'The Creation of a New Underclass in China and its Implications', Centre for the Study of Democracy Paper 05 (10), 2005, pp. 9–10. In my view, the centre is more cautious with applying these 'weapons' to peasants, but the observation that the centre faces conflicts of interest between producing revenue, pacifying local agents and creating social stability hold equally true for rural China. The centre's intervention in the June 2005 Dingzhou riots, and its non-intervention in the Taishi incident, which started in August 2005, illustrates this ambivalence. In Henan's Dingzhou city, an electricity company hired 300 thugs to expel farmers from their land, which had been illegally sold by local authorities and then subsequently occupied by the farmers for two years (!). The Party secretary and the mayor of Dingzhou were recalled after the riots, in which six farmers were killed. P. P. Pan, 'Chinese peasants attacked in land dispute: at least six die as armed thugs assault villagers opposed to seizure of property',

Washington Post, June 15, 2005, p. A12. In Guangdong's Taishi village, peasants recalled the head of the village committee, but withdrew their motion after being intimidated by hired thugs. Roadblocks barred access to the village, and journalists, both Chinese and foreign, were beaten by thugs. T. Luard, 'China village democracy skin deep', *BBC Online*, October 10, 2005. Online. Available HTTP: http://newsvote.bbc.co.uk/mpapps/pagetools/print/news.bbc.co.uk/2/hi/asia-pacific/4319954.stm (accessed on October 10, 2009). In spite of international indignation, the centre justified these actions as 'legal'. Ministry of Foreign Affairs, *Foreign Ministry Spokesperson Kong Quan's Press Conference on October 11, 2005*. Online. Available HTTP: http://www.fmprc.gov.cn/eng/xwfw/s2510/t216357.htm (accessed on October 19, 2009).

175 T. Heberer and W. Taubmann, *Chinas ländliche Gesellschaft im Umbruch. Urbanisierung und sozioökonomischer Wandel auf dem Lande*, Opladen and Wiesbaden: Westdeutscher Verlag, 1998, pp. 272 and 76–9.

176 Zhong, *Local Government and Politics in China*, p. 74.

177 Ibid., p. 75.

178 Ibid., p. 61.

179 L. Chen, *Building Local State Capacity in China*, Chicago: UMI Dissertations, 2004, pp. 67–8.

180 Zhonghua Renmin Gongheguo Guowuyuan ling 125, *Guojia Gongwuyuan zhanxing tiaoli* [Provisional Regulation for Government Employees].

181 Ibid.

182 Chen, *Building Local State Capacity in China*, pp. 183–4.

183 Zhonghua Renmin Gongheguo Guowuyuan ling 125.

184 Zhongfa [1995] No. 4, *Zhonggong zhongyang guanyu yinfa 'dang zheng lingdao ganbu xuanbo rnyong gongzuo zhanxing tiali' de tongzhi* [Notification by the Central Committee of the CCP regarding the printing and distribution of the 'Provisional Regulation for the Selection and Appointment of Party and State Leading Cadres'].

185 Chen, *Building Local State Capacity in China*, p. 73.

186 Ibid., p. 74.

187 Ibid., p. 75.

188 For the complex processes and mechanisms of cadre recruitment see Heberer and Taubmann, *Chinas ländliche Gesellschaft im Umbruch*, pp. 267–9, and Zhong, *Local Government and Politics in China*, pp. 105–19.

189 Chen, *Building Local State Capacity in China*, p. 192.

190 Author's interview, P County, Hebei Province, September 25, 2004.

191 Author's interview, G County, Anhui Province, March 11, 2006.

192 Lam and Perry, 'Service organizations in China: reform and its limits', in P. N.-S. Lee and C. W.-H. Lo (eds), *Remaking China's Public Management*, Westport CT: Quorum Books, 2001, pp. 25–6.

193 C. W.-H. Lo, J. M.-K. Lo and K.-C. Cheung, 'Service organizations in the environmental governance system of the People's Republic of China', in P. N.-S. Lee and C. W.-H. Lo (eds), *Remaking China's Public Management*, p. 53.

194 Chen, *Building Local State Capacity in China*, p. 241.

195 Author's interview, Beijing, October 10, 2004.

196 X. Lü, 'The politics of peasant burden in reform China', *Journal of Peasant Studies* 25 (1), 1997, p. 349.

197 Li, 'Differentiated actors', p. 172.

198 Lü, 'The politics of peasant burden in reform China', pp. 353–4.

199 Funds related to SOE and supervisory departments chiefly consisted of three items: the renovation and reconstruction fund, profit retention, and the major repairs fund. Together, they accounted for around nine-tenths of the total.

200 Of course, some of such fees are also part of the regular budget, though they account only for a minor fraction. In 2005, 'Income from administrative fees' amounted to 107.37 billion yuan, or 3.7 per cent of budgetary income at the national level. The fact that enterprise-related funds made up between 75 per cent and 80 per cent of total EBF during the 1980s might easily mask the fact that administrative funds and fees have been rapidly increasing in relative and absolute terms since 1984, when the decision was made to further reduce state planning. In 1985, such funds increased 63.6 per cent over the previous year, and their average growth rate between 1984 and 1992 stood at 26.3 per cent.

201 Local governments levied surcharges on the industrial and commercial tax, the agricultural and animal husbandry tax, and urban public utilities. See Wong, Heady and Woo, *Fiscal Management and Economic Reform in the People's Republic of China*, p. 236, table 1).

202 Whiting, *Power and Wealth in Rural China*, p. 84.

203 Ibid.

204 C. P. W. Wong and D. J. S. Brean, 'Fiscal dualism in China: gradualist reform and the growth of off-budget revenue', in D. J. S. Brean (ed.), *Taxation in Modern China*, New York and London: Routledge, 1998, pp. 187–208, p. 199.

205 Ibid., p. 197.

206 Whiting, *Power and Wealth in Rural China*, p. 86.

207 Wong, Heady and Woo, *Fiscal Management and Economic Reform in the People's Republic of China*, p. 233.

208 Wong and Brean, 'Fiscal dualism in China', p. 192.

209 F. Gang, 'Market-oriented economic reform and growth of off-budget local public finance', in D. J. S. Brean (ed.) *Taxation in Modern China*, New York and London: Routledge, 1998, p. 219.

210 Ibid., p. 216.

211 Ibid., p. 214; for exactions from private enterprises see Whiting, *Power and Wealth in Rural China*, pp. 96–8.

212 F. Gang, 'Market-oriented economic reform and growth of off-budget local public finance', p. 214.

213 C. Wong, *Financing Local Government in the People's Republic of China*, Hong Kong and New York: Oxford University Press, 1997, p. 201.

214 Wong and Brean, 'Fiscal dualism in China', p. 197, table 8.2.

215 A detailed list of these and other items is to be found in Wong, *Financing Local Government in the People's Republic of China*, p. 202. They are discussed thoroughly by Bernstein and Lü, *Taxation without Representation in rural China*, pp. 50 f.

216 Whiting, *Power and Wealth in Rural China*, p. 85.

217 Ibid., p. 242.

218 Chen and Chun, *Zhongguo nongmin diaocha*, pp. 151–5.

219 Whiting, *Power and Wealth in Rural China*, pp. 91–2.

220 See Bernstein and Lü, *Taxation without Representation in Contemporary Rural China*, pp. 50 f., and F. Li, 'Zhongguo nongmin de zhengzhi xuanze' [Political choices for China's farmers], *Manuscript*, 2004. The head of the finance office of one town confirmed that his government had become less lenient after the RTFR, and that lost revenue was partly compensated for by handing out fines. Author's interview, Anhui Province, M Town, March 16, 2006).

221 'Zhongguo baixing lanpishu: san nong (2)', [White Book of China's Common People: the three rural problems (2)], 2002. Online. Available HTTP: http://news.xinhuanet.com/newscenter/2002–9/09/content_554355.htm (accessed on October 19, 2009).

222 Cited in Wong, *Financing Local Government in the People's Republic of China*, p. 200, table 5.22.
223 Gang, 'Market-oriented economic reform and growth of off-budget local public finance', p. 212.
224 Own calculations from *Songjiang Statistical Yearbooks*, 1991–95, cited in Whiting, *Power and Wealth in rural China*, p. 88.
225 Ibid., p. 285.
226 Own calculation from Anhui Province, RTFR LSG *Anhui Sheng nongcun shuifeigaige qianhou nongmin fudan qingkuang biao* [*Situation of peasant burden before and after the Rural Tax and Fee Reforms in Anhui Province*], 2003, internal reference.
227 S. Wang, 'China's 1994 fiscal reform: an initial assessment', *Asian Survey* 37 (9), 1997, p. 808.
228 While governments of all levels continue to exact unlisted revenues from China's SOEs up to this day, the large-scale restructuring of the state sector has decreased the rent margin.
229 The exclusion of enterprise-related funds from EBF after 1992 made administrative fees the single most important source of non-budgetary revenue. Their growth rate increased rapidly, averaging 40.3 per cent between 1992 and 1996. See *China Yearbook of Finances*, Beijing: Zhongguo tongji chubanshe, various years.
230 Wong and Brean, 'Fiscal dualism in China: gradualist reform and the growth of off-budget finance', p. 192. The term 'off-budget revenue' is used here to refer to the combination of extra-budgetary and extra-systemic revenue. Unfortunately, no consensus exists over the usage of these terms. Some authors (e.g. Gang, 'Market-oriented economic reform and growth of off-budget local public finance') employ 'off-budget revenue' as a synonym for extra-systemic revenue.
231 For the evolution of China's tax system between 1978 and 1993 see K. H. Y. H. Hsiao, *The Government Budget and Fiscal Policy in Mainland China*, Seattle WA: University of Washington Press, 1987; J. Li, *Taxation in the People's Republic of China*, New York: Praeger, 1991; S. Wang and A. Hu, *Zhongguo guojia nengli baogao* [Report on China's state capacity], Shenyang: Liaoning renmin chubanshe, 1993; S. Cong, *Zhongguo caizheng tansuo* [An exploration of China's fiscal system], Shanghai: Shanghai renmin chubanshe, 1994; Wong, Heady and Woo, *Fiscal Management and Economic Reform in the People's Republic of China*, chapter 3; S. Cong, *Zhongguo hongguan caizheng zhengce yanjiu* [China's macro fiscal policy], Shanghai: Shanghai caijing daxue chubanshe, 1998; D. Hauff, *Finanzausgleichspolitik in der Volksrepublik China. Eine empirische Analyse der Haushaltsreform von 1994 im Spannungsfeld zwischen Zentrale und Peripherie* [The politics of budget balancing in the People's Republic of China: an empirical analysis of the 1994 budgetary reform between the opposing poles of the centre and the periphery], Frankfurt am Main etc.: Peter Lang, 2003, chapters B and C.
232 Hauff, *Finanzausgleichspolitik in der Volksrepublik China*, pp. 96–100.
233 This point is elaborated in more detail by T. Wang, 'Wo guo caizheng tizhi gaige ying jieshou gonggong caizheng lilun' [The reform of China's fiscal system must conform to theories of public finance], *Neibu canyu* 47 (740), 2004, pp. 2–8.
234 Hauff, *Finanzausgleichspolitik in der Volksrepublik China*, p. 129.
235 Ibid.
236 See Guofa [1993] No. 85, *Guowuyuan guanyu shixing fenshuizhi caizheng guanli tizhi de jueding* [Decision by the State Council regarding the introduction of a tax separation budget management system].
237 See ibid.
238 World Bank, *China: National Development and Sub-national Finance*.

239 Shanghai, Tianjin, Beijing, Chongqing, Hebei, Shanxi, Inner Mongolia, Shandong, Sichuan, Xinjiang, Zhejiang, Hainan. Ibid.

240 Zhou, *Zhongguo zhengzhi tizhi gaige*, p. 73.

241 X. Zhou, 'Bengkui de bianyuan: Zhongguo nongcun de jiceng zhengfu zhaiwu weiji' [On the brink of collapse: the debt crisis of grass roots governments in rural China]. Online. Available HTTP: http://www.usc.cuhk.edu.hk/wk_wzdetails.asp?id = 3591 (accessed on October 19, 2009).

242 Bernstein and Lü, *Taxation without Representation in Rural China*, pp. 68–72.

243 World Bank, *China: National Development and Sub-national Finance.*

244 *China Finance Yearbook*, various years.

245 Guofa [1996] No. 29, *Guowuyuan guanyu sjiaqiang yusuan wai zijin guanli de jueding* [Decision by the State Council regarding the strengthening of the management of extra-budgetary finances].

246 Hauff, *Finanzausgleichspolitik in der Volksrepublik China*, p. 88; see Guofa [1996] No. 12, *Guowuyuan pizhuan caizhengbu, guojia jihua weiyuanhui, shenjichu, Zhongguo renmin yinhang, jianchabu guanyu qingli jiancha yusuan wai zijin yijian de tongzhi* [Notification by the State Council concerning comment on and transmission of the views of the Ministry of Finance, the State Planning Commission, the Audit Office, the People's Bank of China and the Ministry of Supervision regarding the cleaning up and investigation of extra-budgetary revenue].

247 In absolute terms, however, they have grown by an average 6.7 per cent a year, marking a 66.3 per cent increase between 1997 and 2004. Much of this growth can again be attributed to another change in the statistics in 1998. That year, revenue from 'state-owned enterprises and its governing department' reappeared in EBF statistics, and 'other revenues' were listed as well. The combined value of these items are responsible for 10 per cent of the 13 per cent increase in EBF that year. See *China Finance Yearbook*, various years.

248 Another new category listed in EBF statistics was 'government funds' (*zhengfu xing jijin*), but their value (47.84 billion yuan) roughly equalled a concurrent decrease in 'revenue of administrative and institutional units' (43.24 by yuan). See ibid. This suggests that they had been included in EBF before, though not under a separate listing.

249 W. Shi, *Gong xuan yu zhixuan: xiangzhen renda xuanju zhidu yanjiu* [Popular and direct elections: research on the electoral system for rural People's Congresses], Beijing: Zhongguo shehui kexue chubanshe, 2000.

250 A valuable discussion of existing scholarly work on this issue is that by G. Schubert, 'Village elections in the PRC: a Trojan Horse of democracy?', *Duisburg Working Paper on East Asian Studies* 2002 (19), 2002.

251 Z. Guo and T. P. Bernstein, 'The impact of elections on the village structure of power: relations between the village committees and the Party branches', *Journal of Contemporary China* 13 (39), 2004, pp. 257–75.

252 The VC's legal standing is one of 'mass organizations of self-management at the grass-roots level' (Constitution of the People's Republic of China, Art. 111).

253 A detailed treatment of the relationships between formal and informal institutions can be found in H.-J. Lauth, *Im Schatten demokratischer Legitimität. Informelle Institutionen und politische Partizipation im interkulturellen Demokratienvergleich*, Opladen: Westdeutscher Verlag, 1999, and H.-J. Lauth, 'Informal institutions and democracy', *Democratization* 7 (4), 2000, pp. 21–50.

254 K. Lieberthal and D. M. Lampton, *Bureaucracy, Politics, and Decision Making in post-Mao China*, Berkeley CA: University of California Press, 1992, p. 12.

255 Ibid., p. 8.

256 V. Shue, 'State power and social organization in China', in J. S. Migdal, A. Kohli and V. Shue (eds), *State Power and Social Forces: domination and transformation*

in the Third World, Cambridge and New York: Cambridge University Press, 1994, pp. 65–88.

257 K. Lieberthal, *Governing China: from revolution through reform*, New York: W. W. Norton, 1995.

258 T. Heberer, 'Das politische System der VR China im Prozess des Wandels', in C. Derichs and T. Heberer (eds), *Die Politischen Systeme Ostasiens*, Wiesbaden: VS Verlag, 2008.

259 Ibid., p. 86.

260 J.-h. Chung, *Central Control and Local Discretion in China: leadership and implementation during post-Mao decollectivization*, Oxford and New York: Oxford University Press, 2000, chapter 4.

261 J. Fewsmith, 'Elite politics', in M. Goldman and R. MacFarquhar (eds), *The Paradox of China's post-Mao Reforms*, Cambridge MA: Harvard University Press, 1999 pp. 47–75.

262 Cf. T. Heberer, *Korruption in China. Analyse eines politischen, ökonomischen und sozialen Problems*, Opladen: Westdeutscher Verlag, 1991; X. Lü, *Cadres and Corruption: the organizational involution of the Chinese Communist Party*, Stanford CA: Stanford University Press, 2000.

263 Heberer, 'Das politische System der VR China im Prozess des Wandels', p. 82.

264 Lü, 'The politics of peasant burden in reform China', p. 353.

265 R. Baum and A. Shevchenko, 'The state of the state', in M. Goldman and R. MacFarquhar (eds), *The Paradox of China's post-Mao Reforms*, Cambridge MA: Harvard University Press, 1999, pp. 344–6.

4 The limited agendas: the peasant burden and rural fiscal reforms

266 All three databases allow only restricted access. Funded by a grant by the German Research Foundation (Deutsche Forschungsgemeinschaft, DFG), the German National Library has created a portal that grants scholars affiliated with German research institutions free access to these and other online databases. See www.crossasia.org.

267 http://www.cnbksy.com (accessed October 20, 2009).

268 http://ckrd150.cnki.net/kns50/Navigator.aspx?ID = 1cjbd/ (accessed 20, October 2009). The coverage of earlier years is indicated by an increasing number of articles.

269 Hardly any articles of the *People's Daily* at that time are listed in the National Index, so there is very little overlap between the two databases. Thus the correlation is not caused by the National Index simply listing the *People's Daily* articles.

270 See for example W. Tang, *Public Opinion and Political Change in China*, Stanford CA: Stanford University Press, 2005.

271 Bernstein and Lü, *Taxation without Representation in Contemporary rural China*, p. 42.

272 'Luoshi dang de zhengce, jianqing nongmin fudan' [Carry out the Party's policy, reduce the peasant burden], *Renmin Ribao*, July 5, 1978, p. 2.

273 J. Chen, 'Qianghua gongchandangyuan de zhongcheng yishi' [Strengthen the loyalty of Communist Party members], *Qiushi* (9), 2007, p. 55.

274 Quotes taken from Zhongfa [2000] No. 7.

275 *Renmin Ribao*, 'Luoshi dang de zhengce, jianqing nongmin fudan', p. 2.

276 The 'Gang of Four', four people accused of having inspired the Cultural Revolution, had been in prison since 1976. Obviously, they served as a proxy for lashing out at Mao's leftist successor, Hua Guofeng (see Bernstein and Lü, *Taxation without Representation in rural China*, p. 43).

277 For example, a trial law of village self-governance had been passed in 1987. The law, which had been championed by the former chairman of the National People's Congress, the conservative Peng Zhen, had faced considerable opposition at the local level. One of Peng's main arguments for the law was that it would improve local governance. See Kelliher, 'The Chinese debate over village self-government'. The reporting on the excessive peasant burden the previous year might have served to underscore the necessity of the policy. Due to constraints of space, however, this thought cannot be pursued further.

278 T. P. Bernstein, 'Instability in rural China', in D. L. Shambaugh (ed.), *Is China Unstable?* Armonk NY: M. E. Sharpe, 2000, p. 210.

279 Ibid., p. 199.

280 Ibid., p. 210.

281 For a detailed account of the riots see Bernstein and Lu, 'Taxation without representation: peasants, the central and the local states in reform China', pp. 130–7.

282 Guowuyuan 1990, *Guowuyuan guanyu qieshi jianqing nongmin fudan de tongzhi* [Notification by the State Council concerning the realistic reduction of peasant burden].

283 Zhongfa [1990] No. 16, *Zhonggong zhongyang, guowuyuan guanyu jianjue zhizhi luan shoufei, luan fakuan he ge zhong tanpai de jueding* [Decision of the central committee of the CCP and the State Council concerning the resolute cessation of wanton fee-taking, wanton fining and all kinds of apportionments].

284 Guowuling [1991] No. 92, *Nongmin chengdan feiyong he laowu guanli tiaoli* [Regulation for the management of peasant fiscal burden and compulsory labour].

285 Guobanfa [1992] No. 42, *Guowuyuan bangongting guanyu jin yi bu zuo hao nongmin chengdan feiyong he laowu jiandu guanli gongzuo de tongzhi* [Notification by the general office of the State Council regarding the further improvement of the management and supervision of the peasant fiscal burden and compulsory labour].

286 Bernstein and Lü, 'Taxation without representation: peasants, the central and the local states in reform China', p. 135.

287 Central Committee 1993, *Zhonggong zhongyang bangongting, guowuyuan bangongting guanyu qieshi jianqing nongmin fudan de jinji tongzhi* [Urgent notification by the general office of the Chinese Communist Party's central committee and the general office of the State Council concerning the thorough reduction of the peasant burden].

288 'Shenru modi "xian ting hou qing" zijin duixian. Zhangzhou jianqing nongmin fudan cuoshi shizai. Renshou Xian tingqu qunzhong yijian nongmin renjun fudan xiajiang', [Explore in detail the 'first stop, then clear' method of cashing funds. Zhangzhou's method of reducing peasant burden is dependable. Renshou County listens to the people's opinion, the average peasant burden is decreasing], *Renmin Ribao*, May 30, 1993, p. 2.

289 'Nongmin fudan wenti tuoshi zhi yi. Fudan, na xie heli? Na xie bu heli?' [Penetrating the peasant burden problem, Part 1. Which burdens are legitimate, and which are not?], *Renmin Ribao*, June 6, 1993, p. 2.

290 'Nongmin fudan wenti tuoshi zhi er. Bu neng yong "tanpai" de banfa gao jianshe?' [Penetrating the peasant burden problem, Part 2. Can't apportionment be used to finance construction?], *Renmin Ribao*, June 7, 1993, p. 2.

291 'Nongmin fudan wenti tuoshi zhi san. Zhan zai nongmin de jiaodu xiang wenti' [Penetrating the peasant burden problem, Part 3. Consider problems from the peasants' standpoint], *Renmin Ribao*, June 8, 1993, p. 2.

292 'Nongmin fudan wenti tuoshi zhi si. Jianqing fudan cong he rushou?' [Penetrating the peasant burden problem, Part 4. Where to start reducing burden?], *Renmin Ribao*, June 9, 1993, p. 2.

293 Bernstein, 'Instability in rural China', *Instability in Rural China*, p. 214.
294 Ibid., p. 217.
295 Bernstein and Lü, *Taxation without Representation in rural China*, p. 167.
296 Ibid.
297 B. Chen, '"Reduced and becoming heavy again" call: symptoms and cure. Observing and pondering the bouncing back of farmers' burdens', *Nongmin Ribao*, March 26, 1996, FBIS-CHI-96-119.
298 Z. Zhou, 'Peasant burden has multiple causes', *Fazhi Ribao*, May 21, 1996. p. 7, FBIS-CHI-96-152.
299 Ibid.
300 C. Li, 'Think national, blame local: central–provincial dynamics in the Hu era', *China Leadership Monitor* (17), 2006, pp. 1–24.
301 J.-h. Chung, 'Reappraising central–local relations in Deng's China: decentralization, dilemmas of control, and diluted efforts of reform', in C.-m. Chao and B. Dickson (eds), *Remaking the Chinese State: strategies, society, and security*, London: Routledge, 2001, pp. 61–2; K. O'Brien and L. Li, 'Popular contention and its impact in rural China', *Comparative Political Studies* 38 (3), 2005, pp. 240–4.
302 L. Li, 'Political trust in rural China', *Modern China* 30 (2), 2004, pp. 228–58.
303 These characters' absence from the headlines naturally does not mean that no suggestions were made to tackle the peasant burden. Since the early 1980s, there has been a strand of discussion circling around the heading of 'rural reforms' (*nongcun gaige*). 'Rural reforms' entailed issues as diverse as the improvement of the rural infrastructure, enhancing agricultural productivity and implementing villagers' self-governance. As will be seen in the next section, fiscal reforms were also discussed under this heading, yet not very prominently.
304 This point is made by Maria Edin (M. Edin, 'State capacity and local agent control in China: CCP cadre management from a township perspective', *China Quarterly* 173, 2003, p. 36).
305 Quite possibly, the attention has turned to the fallout of the reform such as issues of rural finances and service provision, or to other pressing problems such as environmental pollution. Due to constraints of space, and because they are not directly related to this thesis, these hypotheses will not be tested.
306 Interview, Anhui Province, Hefei City, August 17, 2004.
307 Chen and Chun, *Zhongguo nongmin diaocha*, pp. 245–9.
308 One *mu* equals approximately one-sixteenth of a hectare.
309 Chen and Chun, *Zhongguo nongmin diaocha*, pp. 251–7.
310 He and Sun (eds), *Zhongguo nongcun shuifei gaige chutan*.
311 Chen and Chun, *Zhongguo nongmin diaocha*, p. 260.
312 Ibid.
313 Ibid., pp. 260–1.
314 Ibid., p. 262.
315 Ibid., pp. 262–3.
316 Ibid., pp. 264–5.
317 Ibid., p. 268.
318 Ibid., pp. 269–74.
319 Quote in A. Zhu and C. Yao, 'Nongcun shuifei gaige shidian yixian zhuizong (yi)' [Tracing the path of the RTFR (Part 1)], *Zhongguo caijing bao*, October 13, 2003.
320 Ibid.
321 Zhang, *Anhui nongcun shuifei gaige shijian yu tansuo*, pp. 63–4.
322 Chen and Chun, *Zhongguo nongmin diaocha*, pp. 274–85.
323 Zhu and Yao, 'Nongcun shuifei gaige shidian yixian zhuizong (yi)'.

324 Ibid.
325 Chen and Chun, *Zhongguo nongmin diaocha*, p. 280.
326 Ibid., p. 266.
327 Ibid., pp. 282–3.
328 Ibid., pp. 283–5.
329 Ibid., pp. 291–2.
330 Ibid., pp. 292–3.
331 Ibid., pp. 294–5.
332 Ibid., pp. 296–7.
333 Ibid., p. 309.
334 I raised these questions during an interview with Li Kunsen, vice-chairman of the standing committee of Anhui's Provincial People's Congress on August 15, 2004, but he refused to address the topic.
335 Present were delegates of the Central Committee's Policy Research Office, the State Council's Research Office and Development Research Centre, the State Political System Reform Commission, the Chinese Academy of Agricultural Studies, the Chinese Academy of Social Sciences, the Ministry of Finance, the Ministry of Agriculture, the Ministry of Domestic Trade. See Chen and Chun, *Zhongguo nongmin diaocha*, p. 309.
336 Zhu and Yao, 'Nongcun shuifei gaige shidian yixian zhuizong (yi)'.
337 Cainongshui (1994) No. 4, *Caizhengbu zhi Anhui Sheng renmin zhengfu guanyu Taihe Xian nongye shuifei gaige wenti de han* [Letter from the Ministry of Finance to the People's Government of Anhui Province regarding the problem of rural tax and fee reform in Taihe County].
338 Chen and Chun, *Zhongguo nongmin diaocha*, p. 310.
339 Ibid., pp. 311–12; Zhang, *Anhui nongcun shuifei gaige shijian yu tansuo*, p. 64.
340 For Anhui Province, see Chen and Chun, *Zhongguo nongmin diaocha*, pp. 302–3. For the national Ministry of Finance see Chen and Chun, *Zhongguo nongmin diaocha*, pp. 321–7.
341 The reform was fully implemented in Taihe Guoyang and Linquan counties, and partly in Mengcheng, Lixin and Yingshang. He and Sun (eds), *Zhongguo nongcun shuifei gaige chutan*.
342 Many of these cases are described in great detail in Chen Guidi and Chun Tao's monograph. As all cases described in the book took place in this region, the title 'Investigation into Northern Anhui's Peasants' would have been more appropriate.
343 Chen and Chun, *Zhongguo nongmin diaocha*, p. 297.
344 He and Sun (eds), *Zhongguo nongcun shuifei gaige chutan*.
345 Author's interview, Anhui Province, Hefei, February 24, 2006.
346 Zhang, *Anhui nongcun shuifei gaige shijian yu tansuo*, p. 386. Unfortunately, county-level data on taxes and fees collected are not available before 1999.
347 Author's interview, Anhui Province, Hefei, February 24, 2006.
348 Ibid.
349 Chen and Chun, *Zhongguo nongmin diaocha*, p. 284.
350 Ibid., pp. 313–14.
351 Taihe County, RTFR Office, *Anhui Taihe Xian nongcun shuifei gaige shijian* [*The experiences with the RTFR in Taihe County, Anhui Province*], cited in He and Sun (eds), *Zhongguo nongcun shuifei gaige chutan*.
352 Zhongfa [1996] No. 13, *Zhonggong zhongyang guowuyuan guanyu qieshi zuo hao nongmin fudan gongzuo de jueding* [Decision by the Central Committee of the CCP and the State Council to do good work in alleviating peasants' financial burden].
353 Chen and Chun, *Zhongguo nongmin diaocha*, pp. 321–7.

354 Zhongfa [1996] No. 13 (italics added).
355 D. Du and J. Liu, 'He Kaiyin xishuo "er ci da baogan"' [He Kaiyin elaborates on the 'Second Household Responsibility System'], *Jingji luntan* 101 (20), 2001.
356 For the construction of 'peasantness' and its political utilization in CCP China see A. Kipnis, 'Within and against peasantness: backwardness and filiality in rural China', *Comparative Studies in Society and History* 37 (1), 1995, pp. 110–35.

5 From innovation networks to competition under hierarchy

357 Author's interview, Anhui Province, Wuhe County, March 7, 2006.
358 Hubei's Yangdang Town, for example, allowed the levying of only the agricultural tax and its surcharge, the agricultural special products tax, the slaughter tax, the township comprehensive fee and village retention, the agricultural development fund, the education fund, the contract and visa fee, the epidemic prevention fee, the fruit farming service fee, the water fee, and compulsory labour demanded in cash. All other levies could be refused by the peasants. The township and village levies, which were not to change for three years, were calculated on the basis of the previous year's income average. The agricultural taxes were computed according to quotas handed down from above, and the contribution to the education fund according to the amount prescribed by the county's 'burden reduction unit' (*jianfu ban*). The water fee was based on the average revenue of the previous three years. A fixed sum was specified also for the rat extinction fee, which was added to the catalogue in 1998. The fruit farming service fee and the insurance fee for stationary treatment of peasants (also added in 1998) were based on voluntary contributions and were contracted with the farmers. Seventy per cent of the fruit farming service fee was to be levied from paddy owners, and 30 per cent from those with dry fields. The revenues from these items were then allocated by the township government among the relevant organisations. Compulsory labour was valued at 2.5 yuan annually per unit, with each farmer having to serve thirty units. In addition, forty yuan were due for each *mu* of distributed land. Altogether, the annual total burden could not exceed 250 yuan per *mu*. In 1998, this amount was reduced to 220 yuan. In contrast to the village and township levies, which were fixed for three years, the 'burden ceiling' could be changed annually. Ma, Huang and Jiang, *Wo guo nongcun shuifei gaige yanjiu*, pp. 41–2.
359 The other three reform measures were investment and credit structure reform, housing system reform and medical system reform. L. Han, 'A commentary on "one, three, five" in mid-year', *Beijing Xinhua Domestic Service*, July 26, 1998, FBIS-CHI-98-210.
360 Ibid.
361 J. Chen, 'Report on the implementation of the 1997 plan for national economic and social development and on a draft 1998 plan for national economic and social development', *Hong Kong Ta Kung Pao*, March 8, 1998. Pp. 8–9, FBIS-CHI-98-067.
362 Zhu, Rongji, *Zhu Rongji zongli zai guowuyuan di yi ci quanti huihi shang de jianghua* zhaiyao [A summary of Premier Zhu Rongji's speech at the first full meeting of the State Council], 2003, internal reference.
363 Chen and Chun, *Zhongguo nongmin diaocha*, p. 347.
364 Jiang, Zemin, *Jiang Zemin zhuxi 1998 nian 9 yue 25 ri zai Anhui kaocha gongzuo shi de jiang hua* [[Party] chairman Jiang Zemin's speech during an investigative visit to Anhui on September 25, 1998], 2003, internal reference.
365 Xiang Huaicheng, Minister of Finance, Chen Yaobang, Minister of Agriculture, and Duan Yingbi, head of the office of the Central Financial and Economic Leading Small Group.

366 CPCCC 1998, *Zhonggong zhongyang guanyu nongye he nongcun gongzuo ruogan zhongda wenti de jueding* [Decision regarding several major problems in agricultural and rural work].
367 Z. Xia, 'Nongcun shuifei gaige zhidu kuangjia jiben queli' [The basic establishment of the RTFR systemic framework], *Xinhua Kuaijiwang*, 2004, obtainable from the author.
368 Zhu and Yao, 'Nongcun shuifei gaige shidian yixian zhuizong (yi)'.
369 Chen and Chun, *Zhongguo nongmin diaocha*, p. 354.
370 Ibid., p. 355.
371 Guangxi Ribao, October 30, 1998, cited in ibid.
372 Zhu and Yao, 'Nongcun shuifei gaige shidian yixian zhuizong (yi)'.
373 Chen and Chun, *Zhongguo nongmin diaocha*, p. 355.
374 'Xiang pledges reform of tax, fee systems to NPC', *Beijing Xinhua*, March 5, 1999, FBIS-CHI-1999-0305.
375 Its first testing ground outside rural China was the reform of transport and vehicle fees. Such a reform had already been attempted in November 1998. The State Council had submitted to the NPC standing committee a legal draft to abolish '216 fees with a total annual revenue of approximately 15.1 billion yuan' related to road and automobile use. The loss would have been compensated for by raising the fuel and automobile purchase taxes. In its first ever show of recalcitrance, however, the NPC standing committee decided not to vote on the proposal, which amounted to rejecting it. The representatives held that the peasants, who rarely used the highways, would not benefit from the fee reductions, but would suffer from increased retail and fuel prices. 'China: NPC committee shelves State Council proposal', *Beijing Xinhua Hong Kong Service*, November 16, 1998, FBIS-CHI-98-322.
376 *Beijing Xinhua*, 'Xiang pledges reform of tax, fee systems to NPC'.
377 Zhu and Yao, 'Nongcun shuifei gaige shidian yixian zhuizong (yi)'.
378 Xia, 'Nongcun shuifei gaige zhidu kuangjia jiben queli'.
379 Zhang, *Anhui nongcun shuifei gaige shijian yu tansuo*, pp. 64–5; author's interview, Anhui Province, Hefei, August 18, 2004.
380 Huaiyuan, Suixi, Laian and Wangjiang.
381 Zhang, *Anhui nongcun shuifei gaige shijian yu tansuo*, p. 64.
382 *Zhonggong zhongyang, guowuyuan guanyu jinxing nongcun shuifei gaige shidian gongzuo de tongzhi.*
383 See ibid. This distinction is maintained throughout all relevant RTFR documents.
384 Ibid.
385 The ASPT, which was implemented only in 1994, ranged from 8 per cent to 31 per cent of the sales value of cash crops. The concrete rates were decided by the localities, as were most of the products that fell into this category. Since it was very costly to determine who planted how many cash crops in any given year, 'average' harvests could not be specified. Despite this problem, local governments were handed down achievement targets by the higher levels. They coped with the problem by levying the ASPT on a per capita basis, or by simply adding it to the Agricultrural Tax. A second ambiguity existed as to when the tax was to be collected. It could be collected along with the Agricultrural Tax, or paid when the products were sold. Document No. 7 decreed that provinces should standardize tax collection by choosing only one of these points.
386 Zhongfa [2000] No. 7.
387 G. Du, 'Jigou gaige chedi, bixu dangzheng fenkai' [For thorough organizational reform, Party and government must be separated], *Jiancha Ribao*, January 16, 2003. Online. Available HTTP: www.jcrb.com/zyw/n6/ca12185.htm (accessed on November 2, 2007).
388 Zhongfa [2000] No. 7.
389 Du, 'Jigou gaige chedi, bixu dangzheng fenkai'.

390 Ibid.
391 Zhongfa [2000] No. 7.
392 Ibid.
393 Zhu, Zhang and Yan, 'Nongcun shuifei gaige shidian he xiangcun guanli tizhi gaige genzong yanjiu baogao', p. 11.
394 Guononggaiban [2002] No. 30, *Guowuyuan Nongcunshuifeigaige Xiaozu Bangongshi guanyu yinfa 'Xiang Huaicheng tongzhi zai quanguo nongcun shuifeigaige shidian gongzuo zuotanhui shang de jiangfa' he 'Du Qingling tongzhi zai quanguo nongcun shuifeigaige shidian gongzuo zuotanhui jieshu shi de jianghua'* [Notification by the office of the State Council's Rural Tax and Fee Reform work group's printing and distribution of 'Comrade Xiang Huaicheng's speech before the national symposium on work in the Rural Tax and Fee Reform pilot areas' and 'Comrade Du Qinglin's closing remarks at the national symposium on work in the Rural Tax and Fee Reforms pilot areas'].
395 F. Xu, 'Quxiao diji shi ru shiye, sheng guan xian jiang you shidian' [Abandoning prefecture-level cities possible, pilot areas for provincial adminstration of counties set up], *Nanfang zhoumo*, September 15, 2005.
396 Zhongfa [2000] No. 7

6 Reform implementation and ad-hoc policy innovation

397 Author's interview, Anhui Province, Hefei, August 17, 2004.
398 Ibid.
399 Wen, Jiabao, *Wen Jiabao fuzongli zai Anhui guancha nongcun shuifei gaige shidian gongzuo shi de zhongyao jianghua* zhaiyao [Summary of Vice-Premier Wen Jiabao's important speech at his inspection of the Rural Tax and Fee Reform Pilot Work in Anhui Province], 2003, internal reference.
400 Wanfa [2000] No. 10, *Zhonggong Anhui shengwei, Anhui sheng renmin zhengfu guanyu zai quan sheng fanwei ne kaizhan nongcun shuifei gaige shidian gongzuo de tongzhi* [Notice of the Communist Party Anhui provincial committee and the Anhui provincial government regarding the beginning of RTFR pilot work in the whole province].
401 Wanbanfa [2000] No. 11, *Zhonggong Anhui shengwei bangongting, Anhui sheng renmin zhengfu bangongting guanyu yinfa Anhui sheng nongcun shuifei gaige youguan peitao wenjian de tongzhi* [Notice of the Communist Party Anhui provincial committee's general office and the Anhui provincial government's general office concerning the printing and distribution of supplementary documents related to the RTFR in Anhui Province].
402 Ibid.
403 F County, RTFR office, *F Xian nongcun shuifei gaige shishi fang'an* [Programme for the implementation of the Rural Tax and Fee Reform in F County], 2002, internal reference. If fees for certain services are not directly levied from the peasants, but are covered by means of the Village Retention receipts, the latter will naturally tend to be higher. Since the *santi wutong* must not exceed 5 per cent of average income, this means that the township comprehensive levy will be comparatively lower.
404 Ibid.
405 This argument makes sense if the county's high level of indebtedness is considered. Had it apportioned a larger part of the *santi wutong* to the townships and financed village-level services directly by irregular fees, it would not have had to incur so many liabilities.
406 F County, RTFR office, *F Xian nongcun shuifei gaige juti caozuo banfa* [Concrete operating measures for the Rural Tax and Fee Reform in F County], 2002, internal reference.

407 Author's interview, W County, Anhui Province, March 8, 2006.
408 Anhui Province, RTFR LSG, *Zhi quan sheng nongmin pengyou de yi feng xin* [A letter to the peasant friends of Anhui Province], 2000, internal reference.
409 Zhongbanfa [2000] No. 30, *Zhonggong zhongyang bangongting, guowuyuan bangongting guanyu shi xian xiang renyuan bianzhi jingjian de yijian* [Opinion of the general office of the Chinese Party's central committee and the general office of the State Council concerning the reduction of staff on the government payroll in cities, counties and townships].
410 Wanfa [2000] No. 15, *Zhonggong Anhui shengwei, Anhui sheng renmin zhengfu guanyu xiangzhen dang zheng jigou gaige de shishi yijan* [Views of Communist Party Anhui provincial committee and Anhui provincial government regarding the implementation of Party and government structural reforms at the township level].
411 Zhou, 'Bengkui de bianyuan: Zhongguo nongcun de jiceng zhengfu zhaiwu weiji'.
412 T. P. Bernstein, 'Can the Peasant Burden Problem be Resolved? A preliminary assessment of recent policy innovations', paper presented at fortieth anniversary reunion conference of the Universities Centre for Chinese Studies, 'The State of Contemporary China', Chinese University of Hong Kong; Zhu, Zhang and Yan, 'Nongcun shuifei gaige shidian he xiangcun guanli tizhi gaige genzong yanjiu baogao', p. 21.
413 Author's interview, Anhui Province, August 17, 2004.
414 Anhui Province, RTFR LSG, *2000 nian quan sheng nongcun shuifei gaige zhuanyi zhifu he buzhu qingkuang biao* [Status table of transfer payments and subsidies related to the Rural Tax and Fee Reform in Anhui Province for the year 2000], 2003, internal reference.
415 Own calculations based on *Anhui Statistical Yearbook 2000*, Hefei: Anhui Tongji Chubanshe, 2000, and Anhui Province, RTFR LSG, *Anhui Sheng nongcun shuifeigaige qianhou nongmin fudan qingkuang biao*. The following chapter will explore the impacts of the reform on Anhui's sixty-one counties in detail.
416 Chen and Chun, *Zhongguo nongmin diaocha*, p. 401.
417 Ibid., pp. 400–1.
418 Gao, 'Nongcun shuifei gaige-redian wenti leng sikao'; Zhu, Zhang and Yan, 'Nongcun shuifei gaige shidian he xiangcun guanli tizhi gaige genzong yanjiu baogao', pp. 14–16.
419 Gao, 'Nongcun shuifei gaige-redian wenti leng sikao'. This and other strategies of local-level shirking and sabotage will be dealt with in more detail below.
420 Li 2007: 95.
421 Xiang, Huaicheng, *Xiang Huaicheng buzhang zai quanguo nongcun shuifei gaige gongzuo zuotanhui shang de jianghua* zhaiyao [Summary of Minister Xiang Huaicheng's speech at the symposium for the nationwide Rural Tax and Fee Reform work], 2000, internal reference.
422 Chen and Chun, *Zhongguo nongmin diaocha*, p. 305.
423 Ibid., pp. 305–6.
424 Xiang, Huaicheng, *Xiang Huaicheng buzhang zai quanguo nongcun shuifei gaige shidian gongzuo huiyi shang zongjie jianghua* zhaiyao [Summary of Minister Xiang Huaicheng's summarizing speech at the conference for the nationwide Rural Tax and Fee Reform pilot work], 2001, internal reference.
425 R. Zhu, 'Jiu jie renda sici huiyi juxing jizhe zhaodaihui: zhu Rongji zongli da Zhong wai jizhe wen', *Renmin Ribao*, March 16, 2001. Online. Available HTTP: http://www.people.com.cn/GB/shizheng/16/20010316/418508.html (accessed on October 1, 2009).
426 Ibid.
427 It is not, however true that the reforms stalled completely in 2001. Jiangsu implemented the RTFR that year, though on a self-financed base.

428 Bernstein, 'Can the peasant burden problem be resolved?'.
429 Li, 'Working for the peasants?', p. 99.
430 Ibid., p. 100.
431 Wanbanmingdian [2000] No. 47, *Zhonggong Anhui sheng bangongting, Anhui sheng renmin zhengfu bangongting 'guanyu renzhen zuo hao nongyeshuishou zhengshou guanli gongzuo' de tongzhi* [General office of the Communist Party Anhui provincial committee, general office of the Anhui provincial government 'Notification regarding the diligent implementation of agricultural tax collection and administration work'].
432 Wanbanfa [2000] No. 14, *Zhonggong Anhui sheng bangongting, Anhui sheng renmin zhengfu bangongting guanyu zhuanfa 'sheng wujia ju, sheng caizheng ting, sheng jianchating, sheng nongwei, sheng zhengfu jiufengban guanyu qingli zhengdun she nong shoufei, qieshi jianqing nongmin fudan de yijian' de tongzhi* [General office of the Communist Party, Anhui provincial committee, general office of the Anhui provincial government, 'Opinions of the Provincial pricing department, finance department, procurate, farmers' comittee and the ethics office regarding the clearance and rectification of agricultural fees and the thorough lessening of farmers' burden'].
433 Wanbanfa [2000] No. 41, *Zhonggong Anhui sheng bangongting 'guanyu zai nongcun shuifei gaige shidian guocheng zhong qieshi zuo hao jianqing nongmin fudan gongzuo de tongzhi'* [General office of the Communist Party Anhui provincial committee, 'Notification concerning the thorough improvement of burden reduction work in the RTFR pilots']; Wannongjian [2001] No. 2, *Anhui Sheng nongmin fudan jiandu guanli lingdao xiaozu 'guanyu jin yi bu guanfan he wanshan nongmin fudan jianduka zhidu de tongzhi'* [Notification by the leading small group for peasant burden supervision and administration of Anhui Province regarding the further standardization and fine-tuning of the peasant burden supervision card system].
434 *Zhonggong Anhui shengwei, Anhui sheng renmin zhengfu guanyu xiangzhen dang zheng jigou gaige de shishi yijan*; Wanfa [2000] No. 16, *Zhonggong Anhui shengwei, Anhui sheng renmin zhengfu guanyu xiangzhen shiye danwei jigou gaige de yijian* [Opinions of the Communist Party Anhui provincial committee and the Anhui provincial government regarding the implementation of the reform of Party and government structures at the township and town levels].
435 Wanfa [2000] No. 22, *Zhonggong Anhui sheng bangongting, Anhui sheng renmin zhengfu bangongting 'guanyu yinfa "guanyu jin yi bu tiaozheng nongcun zhongxiaoxue buju de yijian" deng nongcun shuifei gaige youguan peitao wenjian de tongzhi'* [General office of the Communist Party Anhui provincial committee, general office of the Anhui provincial government, 'Notification regarding the printing and distribution of the "Opinions regarding the further rectification of village primary school arrangements" and other supplementary documents concerning the RTFR'].
436 Wanting [2001] No. 17, *Zhonggong Anhui shengwei bangongting, Anhui sheng renmin zhengfu bangongting guanyu yinfa 'guanyu tiaozheng nonge techan shui youguan zhengce de yijian' de tongzhi* [Notice of the Communist Party Anhui provincial committee's general office and the Anhui provincial government's general office concerning the printing and distribution of the 'Opinions regarding the policies concerning the adjustment of the agricultural special products tax'].
437 Wanbanfa [2000] No. 21, *Zhonggong Anhui shengwei bangongting, Anhui sheng renmin zhengfu bangongting 'guanyu renzhen zuohao wu baohu shenghuo anpai, shoumie diqu nongyeshui jianmian he qieshi jiaqiang nongcun shui, dian deng shengchan jingyingxing feiyong shouqu guanli gongzuo de jinji tongzhi'* [Office of

the Communist Party Anhui provincial committee and the office of the Anhui provincial government's 'Urgent notification regarding the diligent improvement of the five "to be protected"s' living arrangements, agricultural tax reductions in areas stricken by disaster and the thorough strengthening of rural water, electricity and other supplies as well as service fees collection and administration work']; Wanzhengmingdian [2000] No. 35, *Anhui sheng renmin zhengfu 'guanyu jinyibu guifan nongye shui, nongye techan shui ji qi fujia zhengguan gongzou de jinji tongzhi'* [Anhui provincial government's 'Urgent notification regarding further standardization of the collection and administration of the Agricultrural Tax and the agricultural special products tax and their supplementary charges'].

438 Caizong [2001] No. 61, *Caizheng bu, guojia jiwei, nongyebu guanyu jiada zhili xiang nongmin luan shoufei lidu qieshi jianqing nongmin fudan de tongzhi* [Notification by the Ministry of Finance, the State Planning Commission and the Ministry of Agriculture concerning bringing under control with increased force the problem of the collection of illegal fees from the peasants and realistically reducing the peasant burden] was passed one month after Premier Zhu Rongji had visited Anhui (see below). It complained that central policies had not been implemented thoroughly and that the problem of illegal exactions was still 'very serious' in 'several places'. It listed several areas where fee-taking was still rampant, such as education, water and electricity, marriage licences and house-building permits. It urged local government officials to 'vigorously increase efforts to bring under control illegal extortions from the peasants and conscientiously decrease the peasant burden', and demanded that legal fees be standardized and listed on public bulletin boards.

439 Wanfa [2001] No. 6, *Zhonggong Anhui shengwei, Anhui sheng renmin zhengfu guanyu zuo hao 2001 nian quan sheng nongcun shuifei gaige gongzuo de yijiani* [Opinion of the Communist Party anhui provincial committee and the Anhui provincial government regarding theefficient handling of RTFR work in the whole province]; Wanbanfa [2001] No. 7, *Zhonggong Anhui shengwei bangongting, Anhui sheng renmin zhengfu bangongting guanyu yinfa 'guanyu jin yi bu jiaqiang he guifan nongye shuishou zhengshou guanli gongzuo de yijian' deng nongcun shuifei gaige youguan peitao wenjian de tongzhi* [Notice of the Communist Party Anhui provincial committee's general office and the Anhui provincial government's general office concerning the printing and distribution of the 'Opinions regarding the further strengthening and standardization of agricultural tax collection and administration work' and other supplementary documents related to the RTFR]; Wanzhengban [2001] No. 79, *Anhui Sheng renmin zhengfu bangongting guanyu zhuanfa Sheng nongye weiyuanhui guanyu jin yi bu wending nongcun tudi chengbao guanxi wanshan nongye chengbao hetong guanli ruogan yijian de tongzhi* [Distribution by the general office of the government of Anhui Province of the notification by the provincial agricultural committee regarding several opinions concerning the further stabilization of village land contractual relationships and perfecting the administration of the Household Responsibility System]; Wanfa [2002] No. 3, *Zhonggong Anhui shengwei, Anhui sheng renmin zhengfu guanyu zuo hao 2002 nian quan sheng nongcun shuifei gaige gongzuo de yijian* [Opinions of the Communist Party Anhui provincial committee and the Anhui provincial government regarding doing well the province-wide RTFR work in 2002]; Wandanglianfa [2001] No. 6, *Anhui Sheng dang'anju, Anhui Sheng nongcun shuifei gaige lingdao xiaozu bangongshi guanyu yinfa 'Anhui Sheng nongcun shuifei gaige dang'an guanli banfa' de tongzhi* [Notification by the bureau of files and the office of the RTFR LSG regarding the printing and distribution of 'Measures regarding the administration of files related to the RTFR in Anhui Province']; Cainongshui [2002] No. 185, *Anhui Sheng caizhengting guanyu*

tiaozheng maocha, guopin lianglei yingshui pinmu nongye techanshui shuilü de tongzhi [Notification by the Ministry of Finance of Anhui Province regarding the adjustment of the special products tax rate for the two taxable items of semi-made tea and fruits].

440 Wanzhengmingdian [2001] No. 6, *Anhui sheng renmin zhengfu guanyu zhanting xiang nongmin shouqu Huabei pingyuan gaizao he Pishihang Chaohu guanggai xiangmu yiji nongye zonghe kaifa tudi zhili xiangmu huan zhai zijin de jinji tongzhi* [Urgent notification of Anhui provincial government concerning the temporary suspension of the levy on peasants to repay the loans for the Huaibei plain improvement, the Pishihang Chaohu area development project and the comprehensive rural development project]; Wanzhengmingdian [2001] No. 9, *Anhui sheng renmin zhengfu guanyu jiaqiang nongcun yiwu jiaoyu shoufei guanli jianjue zhizhi luan shoufei de tongzhi* [Notification of the Anhui provincial government concerning the strengthening of the administration of rural compulsory education fees and the permanent cessation of irregular fees]; Wanzhengban [2001] No. 60, *Anhui sheng renmin zhengfu bangongting guanyu zhuanfa 'Guowuyuan bangongting guanyu qieshi zu hao dangqian qianjing nongmin fudan gongzuo de tongzhi'* [Notification by the general office of the government of Anhui Province regarding the distribution of the 'Notification by the general office of the State Council concerning the thorough implementation of the present peasant burden reduction work']; Wanbanmingdian [2001] No. 96, *Zhonggong Anhui shengwei, Anhui sheng renmin zhengfu guanyu zhuanfa 'sheng jiwei, shengwei xuanchuanbu, sheng jianchating, sheng xinwenchubanju guanyu zhengdun baokan faxing chixu, qieshi jianqing cunji fudan de ruogan yijian' de tongzhi* [Notice of the Communist Party Anhui provincial committee and the Anhui provincial government regarding distribution of 'Several opinions of the provincial disciplinary committee, propaganda department, procurate, publication department regarding the rectification of the mechanisms for the distribution of publications and the thorough reduction of the village-level burden']; Wanzheng [2001] No. 99, *Anhui sheng renmin zhengfu zhuanfa sheng wujiaju deng liu bumen guanyu zhuanxiang zhili nongmin jianfang shoufei qieshi jianqing nongmin fudan yijian de tongzhi* [Notification by the government of Anhui Province regarding the distribution of the opinions of the provincial pricing and six other departments regarding the special handling of peasant house-building fees and the realistic reduction of the peasant burden]; Wanzhengban [2001] No. 84, *Anhui sheng renmin zhengfu bangongting zhuanfa sheng wujiaju deng wu bumen guanyu jiaqiang nongcun hunying dengji, jihua shengyu deng shoufei guanli yijian de tongzhi* [Notification by the general office of the government of Anhui Province regarding the distribution of the opinions of the provincial pricing and five other departments regarding the improved management of the marriage registration fee and the birth control fee]; Wanzhengmingdian [2001] No. 20, *Anhui sheng renmin zhengfu guanyu jin yi bu luoshi nongcun shuifei gaige zhengce zuo hao wending nongmin fudan gongzuo de jinji tongzhi* [Urgent Notification by the Government of Anhui Province regarding the further implementation of the RTFR policies and the improved stabilization of peasant burdens]; Wanbanfa (2002) No. 10, *Anhui Sheng renmin zhengfu bangongting zhuanfa guowuyuan bangongting zhuanfa nongyebu deng bumen guanyu 2002 nian jianqing nongmin fudan gobgzuo yijian de tongzhi* [Notification by the general office of the government of Anhui Province regarding the distribution of the notification by the office of the State Council regarding the distribution of the opinion of the Ministry of Agriculture and other ministries regarding peasant burden reduction in 2002]; Jiaojian [2001] No. 3, *Anhui sheng jiaoyuting, Anhui sheng jianchating, Zhongguo baoxian jiandu guanli weiyuanhui Hefei chipaiyuan banshichu 'guanyu yanjin nongcun zhongxiao*

xuexiao daiban xuesheng baoxian de tongzhi' [Notification by the Ministry of Education, the procuratorial department of Anhui Province, and the Hefei representative of the committee for administration and supervision for China's insurances regarding the strict illegality of rural primary and middle schools handling student insurances]; Wannongjian [2001] No. 3, *Anhui sheng nongmin fudan jiandu guanli lingdao xiaozu 'guanyu queding jianqing nongmin fudan zhongdian gongzuo zeren danwei de tongzhi'* [Notification by the leading small group for peasant burden supervision and administration of Anhui Province regarding the confirmation of the units with special responsibilities in peasant burden reduction]; Shengjiuban [2002] No. 1, *Anhui sheng renmin zhengfu jiuzheng hangye buzheng zhi feng bangongshi 'guanyu chengli sheng she nong anjian chachu bangongshi de tongzhi'* [Notification by the bureau for government ethics of Anhui Province 'regarding the establishment of an office for the investigation and treatment of rural cases occurring in Anhui Province']; Caiji [2001] No. 560, *AnhuisSheng caizhengting guanyu zhanting xiang nongmin fentan he shouju jiaqiang guangai nongye xiangmu huandai zijin de jinji tongzhi* [Urgent notification by the Ministry of Finance of Anhui Province regarding the temporary cessation of apportioning among and obtaining funds from the peasants for paying back loans obtained for improving rural irrigation projects].

441 Wanbanfa [2001] No. 18, *Zhonggong Anhui shengwei bangongting, Anhui sheng renmin zhengfu bangongting 'guanyu jiang nongcun chuzhong, xiaoxue laoshi gongzi shangshou dao xian guanli de tongzhi'* [General office of the Anhui Provincial Communist Party committee and the general office of the Anhui provincial government, 'Notification regarding the transfer of rural middle and primary schoolteachers' salaries administration to the county level'].

442 Wanbanfa [2001] No. 19, *Zhonggong Anhui shengwei bangongting, Anhui sheng renmin zhengfu bangongting guanyu yinfa 'guanyu fangzhi xiang cun liang ji fasheng xin de buliang zhaiwu de yijian' deng nongcun shuifei gaige youguan peitao wenjian de tongzhi* [Notice of the Communist Party Anhui provincial committee's general office and the Anhui provincial government's general office concerning the printing and distribution of the 'Opinions regarding the prevention of new bad credits at the township and village levels' and other supplementary documents related to the RTFR]; Wanzhengban [2002] No. 5, *Anhui sheng renmin zhengfu bangongting guanyu jiji wentuo de huajie cunji zhaiwu de yijian* [Opinions by the general office of the government of Anhui Province regarding the proactive and stable dissolution of village-level debts].

443 Wanbanfa [2001] No. 7.

444 In 2005, roughly 3.5 million households received such subsidies. *China Statistical Yearbook* 2004.

445 The 'five guarantees households' included 19 million persons received income subsides. Ibid. However, the payment of lowest cost-of-living subsidies (*dibao*) is being extended to the rural areas. In 2007, 30 million people alone received rural *dibao* payments.

446 Wanbanfa [2001] No. 7.

447 *2001 nian Anhui nianjian* [Anhui Yearbook 2001], Hefei: Anhui Tongji Chubanshe, 2001, p. 31.

448 Author's interview, Anhui Province, Hefei, August 17, 2004.

449 Author's interview, Anhui Province, W County, March 8, 2006.

450 Caiyu [2000] No. 134, *Caizhengbu guanyu xiafa 'gaige he wanshan nongcun shuifeigaige shidianxian, xiang caizheng guanli tizhi de zhidaoxing yijian'* [Notification by the Ministry of Finance concerning the submission of the 'Guiding opinions regarding the reform and perfection of the county- and township-level financial administration systems in the Rural Tax and Fee Reforms trial area'].

451 Caiyu [2002] No. 468, *Caizhengbu guanyu yinfa 'ongcun shuifei gaige zhongyang dui difang zhuanyi zhifu zhanxing banfa' de tongzhi* [Notification by the Ministry of Finance concerning the printing and distribution of the 'Temporary measures regarding the transfer payments from the central government to the local governments in the Rural Tax and Fee Reform program'].
452 P. Zhang (ed.), *Anhui nongcun shuifei gaige shijian yu tansuo* [Practice and exploration of Anhui's Rural Tax and Fee Reforms], Beijing: Dangdai Chubanshe, 2001.
453 L. Wu, 'Nongcun shuifei gaige dui xiangzhen caizheng de yingxiang ji qi houguo. yi Anhui sheng wei li' [The effects of the Rural Tax and Fee Reforms on township and town finances and their consequences, taking Anhui Province as an example]. Online. Available HTTP: http://www.usc.cuhk.edu.hk/wk_wzdetails.asp?id = 2114 (accessed on June 23, 2009).
454 Author's interview, Beijing, October 10, 2004.
455 Author's interview, Anhui Province, W County, March 8, 2006.
456 Ibid.
457 Ibid.
458 For the English translation of the record of the meeting see S. Liu, 'Compulsory education must on no account be weakened: sidelights on Premier Zhu Rongji's inspection of compulsory education in the Anhui rural areas', *Beijing Xinhua Domestic Service*, July 22, 2001, FBIS-CHI-2001-0722.
459 Wanbanfa [2001] No. 18.
460 Author's interview, Anhui Province, Hefei, August 17, 2004.
461 Anhui Province, RTFR LSG, *2001 nian quan sheng nongcun shuifei gaige zhuanyi zhifu he buzhu qingkuang biao* [Status table of transfer payments and subsidies related to the Rural Tax and Fee Reform in Anhui Province for the year 2001], 2002, internal reference.
462 Author's interview, Beijing, March 28, 2006.
463 Guobanfa [2002] No. 25.
464 In mathematical terms, the following formula was utilized: Local subsidy coefficient = ((Share of agricultural taxes and fees in local budget / National average of share of agricultural taxes and fees in local budgets * Relevance factor) + (Share of personnel and administrative expenses in local expenditures / National average of share of personnel and administrative expenses in local expenditures * Relevance factor)) * Central government fiscal burden coefficient.
465 Own calculations based on *Anhui Statistical Yearbook 2002*, Hefei: Anhui Tongji Chubanshe, 2002, and *Anhui Statistical Yearbook 2003*, Hefei: Anhui Tongji Chubanshe, 2003.
466 Li, 'Working for the peasants?', p. 101.
467 Chung, *Central Control and Local Discretion in China: leadership and implementation during post-Mao decollectivization*, p. 176.
468 Author's interview, Anhui Province, W County, March 8, 2006.
469 Author's interview, Anhui Province, W County, March 8, 2006.
470 Author's interview, Anhui Province, He County, August 16, 2004.
471 He Xian zhengfu, *He Xian xiangzhen caizheng guanli fangshi gaige gongzuo qingkuang huibao* [Work status report on the township-level fiscal administration reform in He County], internal reference.
472 Ibid.
473 Author's interview, Anhui Province, He County, August 16, 2004.
474 Author's interview, Anhui Province, He County, August 16, 2004.
475 Author's interview, Anhui Province, Hefei, August 17, 2004.
476 Author's interview, Hebei Province, P County, September 23, 2004.
477 Author's interview, Hebei Province, P County, September 24, 2004.
478 Author's interview, Hebei Province, P County, September 24, 2004.

479 Zhu, Zhang and Yan, 'Nongcun shuifei gaige shidian he xiangcun guanli tizhi gaige genzong yanjiu baogao', pp. 16–18.
480 Zhang, 'Shuifei gaige hou de de "bishui xiaoying" '; see also D. Zhang, ' "Bi shui" hai zai jixu' [Tax enforcement still continues], *Gaige neican* (31), 2003, pp. 20–4.
481 Guononggai [2002] No. 10, *Guowuyuan nongcun shuifei gaige gongzuo xiaozu guanyu nongcun shuifei gaige shidian diqu zhanting qingshou nongmin shuifei weiqian gongzuo youguan wenti de jinji tongzhi* [Urgent notification by the State Council's Rural Tax and Fee Reform work group concerning questions regarding the temporary cessation of settling small tax and fee arrears with farmers in the Rural Tax and Fee Reform pilot areas].
482 C. Li, 'Nongmin de pinkun: tudi, quanli, ziyou fazhan' [Peasant poverty: land, power, development of freedom], in Y. Zhou (ed.) *Nongmin! Nongmin!*, Guangzhou: Huacheng chubanshe, 2004.
483 Author's interview, Beijing, October 10, 2004.
484 H. Chang, 'Xiang quan zhi bian' [Township capacities are changing], *Caijing Magazine*, October 9, 2004. Online. Available HTTP: http://finance.sina.com.cn/g/20041009/11361067097.shtml (accessed on October 10, 2009).
485 X. Gao and Y. Yang, 'Cong nongcun shuifei gaige kan xiangzhen zhengfu gonggong chanpin de gongji' [Looking at the influence of the rural tax and fee reforms on the supply of public goods by township and town governments], *Gaige neican* 2003 (18), 2003, pp. 36–8; Gao, 'Nongcun shuifei gaige-redian wenti leng sikao'; Q. Yu, 'Nongcun shuifei gaige de diaojiu ji xin jucuo' [Survey and new measures on the reform in tax and fees in the countryside], *Huabei gongxueyuan xuebao* (*sheke ban*) 19 (3), 2003, pp. 84–6; K. Z. Jia, Quanhou, 'Jiefu zhihou: nongcun shuifei gaige dai jiejue de wenti ji duice tantao' [After burden reduction: problems of the Rural Tax and Fee Reform waiting to be solved and an exploration of solutions], *Caizheng Yanjiu* (05), 2002, pp. 5–8.
486 Bernstein, 'Can the peasant burden problem be resolved?'.
487 Author's interview, Anhui Province, G County, March 14, 2006.
488 He Xian zhengfu, *He Xian xiangzhen caizheng guanli fangshi gaige gongzuo qingkuang huibao*, 2004, internal reference.
489 Author's interview, Anhui Province, Wuhe County, March 19, 2006.
490 Author's interview, Anhui Province, Wuhe County, March 19, 2006.
491 H. Chang, '2005: Zhongguo nongye nongcun nongmin xiandaihua zhi shang' [2005: the early death of the modernization of agriculture, villages and farmers], *Caijing Magazine internet edition*, February 4, 2005. Online. Available HTTP: http://finance.sina.com.cn/g/20050204/15321349921.shtml (accessed on October 19, 2009).
492 'Jigou jingjian kaizhi zengjia: zhongguo xiangzhen gaige mianlin jiannan xuanze' [Simplification of the organisational structure or increasing expenses: the reform of China's township-level governments faces difficult choices], *Zhongguo Jingying Bao*, February 19, 2005. Online. Available HTTP: http://finance.sina.com.cn/review/20050219/09571368307.shtml (accessed on October 19, 2009).
493 Author's interview, Anhui Province, W village, March 16, 2006.
494 Wanfa (2004) No. 9, *Anhui sheng renmin zhengfu guanyu quanmian quxiao nongye shui fujia de tongzhi* [Notice by the government of Anhui Province regarding the complete abolition of the supplement to the Agricultrural Tax].
495 Author's interview, Anhui Province, W village, March 16, 2006.
496 Ibid.
497 Author's interview, Anhui Province, W village, March 17, 2006.
498 Author's interview, Anhui Province, W village, March 16, 2006.
499 Author's interview, Anhui Province, W village, March 15, 2006.
500 Author's interview, Anhui Province, W village, March 16, 2006.

501 Anhui Province, RTFR LSG, *Anhui Sheng nongcun shuifeigaige qianhou nongmin fudan qingkuang biao.*
502 Zhu, Zhang and Yan, 'Nongcun shuifei gaige shidian he xiangcun guanli tizhi gaige genzong yanjiu baogao', p. 10.
503 Xiang, Huaicheng, *Xiang Huaicheng tongzhi zai quanguo nongcun shuifei gaige shidian gongzuo zuotanhui shang de jianghua* [Comrade Xiang Huaicheng's speech at the national Rural Tax and Fee Reform pilot work symposium], 2002, internal reference.
504 Author's interview, Shandong Province, Y County, September 16, 2004.
505 Anhui Province, RTFR LSG, *Ge shi nongcun shuifei gaige shidian gongzuo zongshu* [Summary of each city's Rural Tax and Fee Reform pilot work], 2003, internal reference.
506 Guononggai [2002] No. 9, *Guowuyuan nongcun shuifei gaige gongzuo xiaozu, guowuyuan jiuzheng hangye buzheng zhi feng bangongshi guanyu qieshi jiaqiang nongcun shuifei gaige shidian gongzuo de jinji tongzhi* [Urgent notification by the State Council's Rural Tax and Fee Reform working group and the government ethics office concerning realistic strengthening of work in the Rural Tax and Fee Reform pilot areas].
507 Xiang, Huaicheng, *Xiang Huaicheng tongzhi zai quanguo nongcun shuifei gaige shidian gongzuo zuotanhui shang de jianghua.*
508 Caishuifa [2002] No. 139, *Guojia shuiwu zongju guanyu qieshi luoshi zhonggong zhongyang bangongting guowuyuan bangongting 'guanyu dui sheji nongmin fudan an (shi) jian shixing zeren zhuijiu de zhanxing banfa' de tongzhi* [Notification by the national tax administrations's general office concerning the thorough implementation of the general office of the Chinese Communist Party and the general office of the State Council's 'Temporary measures regarding the conduction of investigations of the responsibility for cases and incidents involving the peasant burden'].
509 Guononggai [2002] No. 10.
510 'Chinese farmers to bid farewell to agricultural tax', Xinhua News Agency, March 8, 2004. Online. Available HTTP: http://www2.chinadaily.com.cn/chinagate/doc/2004–03/08/content_312773.htm (accessed on October 19, 2009).
511 State Council general office, *Report on the Work of the Government 2005.* Online. Available HTTP: http://www.gov.cn/english/official/2005–7/29/content_18351.htm (accessed on June 23, 2009).
512 'All agricultural taxes to be scrapped in 2006', *Chinanet*, March 5, 2005. Online. Available HTTP: http://www.china.org.cn/english/China/121852.htm (accessed on October 19, 2009).
513 Guofa [2004] No. 21, *Guowuyuan guanyu zuohao 2004 nian shenhua nongcun shuifei gaige shidian gongzuo de tongzhi* [Notice by the State Council regarding doing well the intensivation of RTFR pilot work in 2004].
514 Guobanfa [2004] No. 68, *Guowuyuan bangongting guanyu chengli shenhua nongcun shuifei gaige shidian gongzuo zhidaoxing wenjian qicao xiaozu he zhuanti xiaozu de tongzhi* [Notice by the general office of the State Council regarding the establishment of authoring and expert groupd for the drafting of guideline documents related to the intensivation of RTFR pilot work].
515 Zhongfa [2006] No. 1, *Zhonggong zhongyang guowuyuan guanyu jiji fazhan xiandai nongye zashi tuijin shehui zhiyi xin nongcun jianshe de ruogan yijian* [Several opinions of the central party committee of the Communist Party and the State Council regarding the proactive development of a modern agriculture and solidly promoting the construction of a New Socialist Countryside].
516 Guobanhan [2006] No. 63, *Guowuyuan bangongting guanyu tongyi guowuyuan nongcun shuifei gaige gongzuo xiaozu ji qi bangongshi bianming de fuhan*

[Affirmative reply by the general office of the State Council to the State Council RTFR working group and its office regarding its change of name].

517 State Council general office, *Report on the Work of the Government 2005*.

518 Ibid.

519 L. Fan, 'Anhui: cong shuifei gaige dao nongcun zonghe gaige' [Anhui: from the tax and fee reforms to comprehensive village reforms], *Sinanet*, March 6, 2005. Online. Available HTTP: http://finance.sina.com.cn/g/20050306/17591407141.shtml (accessed on October 20, 2009)

520 Y. Li, 'Hubei Sheng xiangzhen zonghe peitao gaige dui nongcun hexie shehui goujian de jiji yingxiang' [The positive influence of Hubei Province's township and town comprehensive complementary reforms on building a harmonious society in the countryside], *Guojia Xingzheng Xueyuan*. Online. Available HTTP: www.nsa.gov.cn/download/liyoucai.doc (accessed on October 23, 2005).

521 Guofa [2006] No. 34, *Guowuyuan guanyu zuohao nongcun zonghe gaige gongzuo youguan wenti de tongzhi* [Notice by the State Council regarding several problems relating to handling well the work in the rural comprehensive reform].

522 Own calculations from *China Statistical Yearbook*, various years.

523 Zhu, Zhang and Yan, 'Nongcun shuifei gaige shidian he xiangcun guanli tizhi gaige genzong yanjiu baogao', pp. 20–1. This was confirmed by all my interviewees in official positions.

524 Author's interviews, Anhui Province, F County, August 14, and Hefei, August 17, 2004.

525 Zhu, Zhang and Yan, 'Nongcun shuifei gaige shidian he xiangcun guanli tizhi gaige genzong yanjiu baogao', p. 14.

526 Q. Lu, 'Chen Xiwen: Xin nongcun jianshe huiji yi wan nongmin. Zhongyang gei si xiang buzhu' [Chen Xiwen: the consruction of the New Socialist Countryside benefits hundreds of millions of peasants. The centre gives four kinds of subsidies], *Renmin Ribao Haiwai ban*, May 22, 2007. Online. Available HTTP: http://www.gov.cn/jrzg/2007–{05/22}/content_621803.htm (accessed on October 20, 2009).

527 Ibid.

528 Ibid.

529 Li, 'Working for the peasants?', p. 96.

7 Impacts and outcomes: the case of Anhui Province

530 GDP per capita, percentage of GDP of the three sectors in the local economy; primary, secondary and tertiary GDP per capita.

531 Peasant household size, the amount of land each person cultivates, the percentage of persons of the total population not engaged in agriculture, the percentage of wages in peasant income and the acreage each person sows; employment of agricultural machinery.

532 Access to electricity, access to drinking water, connection to roads.

533 Bernstein and Lü, *Taxation without Representation in rural China*, p. 8.

534 N. Heerink, M. Kuiper and X. Shi, 'China's new rural income support policy: impacts on grain production and rural income inequality', *China and World Economy* 14 (6), 2006, pp. 58–69.

535 Linquan, Funan, Yingshang and Lixin County. See *Anhui Statistical Yearbook 2006*, Hefei: Anhui Tongji Chubanshe, 2006.

536 Unfortunately, health expenditures were not a separate item in Anhui's county-level statistics before 2003, so comparable figures for the year 2000 are not available.

537 The change in expenditure structure denotes the percentage change of each expenditure item in total expenditures calculated between 2000 and 2003 for the first phase, and 2003 and 2005 for the second.

538 Seemingly, capital construction has become a focus item as well. However, the correlation is mainly caused by four extremely large values. If these are omitted, the relationship ceases to exist.

539 Chen, *Zhongguo xian xiang caizheng yu nongmin zengshou wenti yanjiu*, p. 120.

8 Why the 'Third Rural Revolution' failed

540 G. Sartori, 'Concept misformation in comparative politics', *American Political Science Review* 64 (4), 1970, pp. 1033–53.

541 Hutchcroft, 'Centralization and decentralization in administration and politics'.

542 Heilmann, 'Policy Experimentation in China's Economic Rise'.

543 J. Xiang, *Zouchu 'Huang Zongxi dinglü' de guaiquan: zhongguo nongcun shuifei gaige de diaocha yu yanjiu* [Breaking out of the vicious circle of Huang Zongxi's law: investigation and research on the RTFR in China], Xi'an: Xibei daxue chubanshe, 2004.

Bibliography

Alpermann, B., 'Economic Transformation and State Capacity: the case of the Chinese cotton sector', unpublished dissertation, University of Cologne, 2006

Anhui province, RTFR LSG, *Zhi quan sheng nongmin pengyou de yi feng xin* [A letter to the peasant friends of Anhui province], April 26, 2000, internal reference, 2000

—— *2000 nian quan sheng nongcun shuifei gaige zhuanyi zhifu he buzhu qingkuang biao* [Status table of transfer payments and subsidies related to the Rural Tax and Fee Reform in Anhui province for the year 2000], internal reference, 2001

—— *2001 nian quan sheng nongcun shuifei gaige zhuanyi zhifu he buzhu qingkuang biao* [Status table of transfer payments and subsidies related to the Rural Tax and Fee Reform in Anhui province for the year 2001], internal reference, 2002

—— *Anhui Sheng nongcun shuifeigaige qianhou nongmin fudan qingkuang biao* [Situation of peasant burden before and after the Rural Tax and Fee Reforms in Anhui province], internal reference, 2003

—— *Ge shi nongcun shuifei gaige shidian gongzuo zongshu* [Summary of each city's Rural Tax and Fee Reform pilot work], internal reference, 2003

Baldi, B., 'Beyond the federal–unitary dichotomy', *Institute of Governmental Studies Working Paper* 99 (7), 1999

Bardhan, P., 'Decentralization of governance and development', *Journal of Economic Perspectives* 16 (4), 2002, pp. 185–205

Bardhan, P. K. and Mookherjee, D., 'The rise of local governments: an overview', in P. K. Bardhan and D. Mookherjee (eds), *Decentralization and Local Governance in Developing Countries: a comparative perspective*, Cambridge MA: MIT Press, 2006, pp. 1–52

Baum, R. and Shevchenko, A., 'The state of the state', in M. Goldman and R. MacFarquhar (eds), *The Paradox of China's post-Mao Reforms*, Cambridge MA: Harvard University Press, 1999, pp. 333–60

Bernstein, T. P., 'Instability in rural China', in D. L. Shambaugh (ed.), *Is China Unstable? Assessing the factors*, Armonk NY: M. E. Sharpe, 2000, pp. 95–111

—— 'Can the Peasant Burden Problem be Resolved? A preliminary assessment of recent policy innovations', paper presented at fortieth Anniversary Reunion Conference of the Universities Centre for Chinese Studies, 'The State of Contemporary China', Chinese University of Hong Kong, 2004

Bernstein, T. P. and Lü, X., 'Taxation without representation: peasants, the central and the local states in reform China', *China Quarterly* 163, 2000, pp. 742–63

Bernstein, T. P. and Lü, X., *Taxation without Representation in Rural China*, Cambridge and New York: Cambridge University Press, 2003

Brandt, L., Liu, H., Rozelle, S. and Zhang, L., *China. Rural Public Finance: the township perspective*, Annex 8: World Bank Report: Fiscal Reform and Role of the Township, Washington, DC: World Bank, 2006

Brehm, J. and Gates, S., *Working, Shirking, and Sabotage: bureaucratic response to a democratic public*, Ann Arbor MI: University of Michigan Press, 1997

Brodsgaard, K. E., 'Institutional reform and the bianzhi system in China', *China Quarterly* 170, 2002, pp. 361–86

Burns, J. P., '"Downsizing" the Chinese state: government retrenchment in the 1990s', *China Quarterly* 175, 2003, pp. 775–802

Byrd, W. A. and Lin, Q., *China's Rural Industry: structure, development, and reform*, Oxford and New York: Oxford University Press, 1990

Cai, Y., 'Between state and peasant: local cadres and statistical reporting in rural China', *China Quarterly* 163, 2000, pp. 783–805

Caiji [2001] No. 560, *Anhui Sheng caizhengting guanyu zhanting xiang nongmin fentan he shouju jiaqiang guangai nongye xiangmu huandai zijin de jinji tongzhi* [Urgent notification by the Ministry of Finance of Anhui province regarding the temporary cessation of apportioning among and obtaining funds from the peasants for paying back loans obtained for improving rural irrigation projects]

Cainongshui [1994] No. 4, *Caizhengbu zhi Anhui sheng renmin zhengfu guanyu Taihe xian nongye shuifei gaige wenti de han* [Letter from the Ministry of Finance to the people's government of Anhui province regarding the problem of a Rural Tax and Fee Reform in Taihe County]

—— [2002] No. 185, *Anhui Sheng caizhengting guanyu tiaozheng maocha, guopin lianglei yingshui pinmu nongye techanshui shuilü de tongzhi* [Notification by the Ministry of Finance of Anhui province regarding the adjustment of the special products tax rate for the two taxable items of semi-made tea and fruits]

Caishuifa [2002] No. 139, *Guojia shuiwu zongju guanyu qieshi luoshi zhonggong zhongyang bangongting guowuyuan bangongting 'guanyu dui sheji nongmin fudan an (shi) jian shixing zeren zhuijiu de zhanxing banfa' de tongzhi* [Notification by the National Tax Administration's general office concerning the thorough implementation of the general office of the Chinese Communist Party and the general office of the State Council's 'Temporary measures regarding the conduction of investigations of the responsibility for cases and incidents involving peasant burden']

Caishuizi [2000] No. 67, *Caizhengbu, Guojia shuiwu zongju guanyu nongcun shuifeigaige shidian diqu nongye techanshui zhengce de tongzhi* [Notification by the Ministry of Finance and the National Tax Administration general office concerning the policies towards the special products tax in Rural Tax and Fee Reform trial areas]

—— [2000] No. 71, *Caizheng bu, guojia shuiwu zongju guanyu nongcun shuifei gaige shidian diqu quxiao tuzaishui de tongzhi* [Notification by the Ministry of Finance and the State Tax General Department regarding the abolition of the slaughter tax in the Rural Tax and Fee Reform pilot regions]

Caiyu [2000] No. 134, *Caizhengbu guanyu xiafa 'gaige he wanshan nongcun shuifeigaige shidianxian, xiang caizheng guanli tizhi de zhidaoxing yijian'* [Notification by the Ministry of Finance concerning the submission of the 'Guiding opinions regarding the reform and perfection of the county- and township-level financial administration systems in the Rural Tax and Fee Reform pilot area']

—— [2002] No. 468, *Caizhengbu guanyu yinfa 'nongcun shuifei gaige zhongyang dui difang zhuanyi zhifu zhanxing banfa' de tongzhi* [Notification by the Ministry of

Finance concerning the printing and distribution of the 'temporary measures regarding the transfer payments from the central government to the local governments in the Rural Tax and Fee Reform programme']

Caizong [2001] No. 61, *Caizheng bu, guojia jiwei, nongyebu guanyu jiada zhili xiang nongmin luan shoufei lidu qieshi jianqing nongmin fudan de tongzhi* [Notification by the Ministry of Finance, the State Planning Commission and the Ministry of Agriculture concerning bringing under control with increased force the problem of the collection of illegal fees from the peasants and realistically reducing the peasant burden]

Central Committee 1993, *Zhonggong zhongyang bangongting, guowuyuan bangongting guanyu qieshi jianqing nongmin fudan de jinji tongzhi* [Urgent notification by the general office of the Chinese Communist Party's central committee and the general office of the State Council concerning the thorough reduction of the peasant burden]

Chang, C., McCall, B. P. and Wang, Y., 'Incentive contracting versus ownership reforms: evidence from China's township and village enterprises', *Journal of Comparative Economics* 31 (3), 2003, pp. 414–28

Chang, H., 'Xiang quan zhi bian' [Township capacities are changing], *Caijing Magazine*, October 9, 2004. Online. Available HTTP: http://finance.sina.com.cn/g/20041009/11361067097.shtml (accessed on November 2, 2009)

—— '2005: zhongguo nongye nongcun nongmin xiandaihua zhi shang' [2005: the early death of the modernization of agriculture, villages and farmers], *Caijing Magazine internet edition*, February 4, 2005. Online. Available HTTP: http://finance.sina.com.cn/g/20050204/15321349921.shtml (accessed on June 23, 2009)

Chen, B., '"Reduced and becoming heavy again" call: symptoms and cure. Observing and pondering the bouncing back of farmers' burdens', *Nongmin Ribao*, March 26, 1996, FBIS-CHI-96-119

Chen, G. and Chun, T., *Zhongguo nongmin diaocha* [An inquiry into the situation of China's farmers], Beijing: Renmin wenxue chubanshi, 2004

—— *Will the Boat sink the Water? The life of China's peasants*, New York: Public Affairs, 2006

Chen, J., 'Report on the implementation of the 1997 plan for national economic and social development and on a draft 1998 plan for national economic and social development', *Hong Kong Ta Kung Pao*, March 8, 1998, pp. 8–9, FBIS-CHI-98-067

—— 'Qianghua gongchandangyuan de zhongcheng yishi' [Strengthen the loyalty of Communist Party members], *Qiushi* (9), 2007, p. 55

Chen, L., *Building Local State Capacity in China*, Chicago: UMI Dissertations, 2004

Chen, X., *Zhongguo xian xiang caizheng yu nongmin zengshou wenti yanjiu* [Research on public finances in China's counties and townships and the problem of increasing peasant income], Taiyuan: Shanxi jingji chubanshe, 2003

Chen, X., Han, J. and Zhao, Y., *Zhongguo nongcun gonggong caizheng zhidu: lilun, zhengce, shizheng yanjiu* [Public finance in rural China: theory, policies, empirical research], Beijing Shi: Zhongguo fazhan chubanshe, 2005

Chung, J.-h., *Central Control and Local Discretion in China: leadership and implementation during post-Mao decollectivization*, Oxford and New York: Oxford University Press, 2000

—— 'Reappraising central–local relations in Deng's China: decentralization, dilemmas of control, and diluted efforts of reform', in C.-m. Chao and B. Dickson (eds), *Remaking the Chinese State: strategies, society, and security*, London: Routledge, 2001, pp. 47–75

CIA, *The CIA World Factbook on CD-ROM*, Minneapolis MN: Quanta Press, 2007
Cong, S., *Zhongguo caizheng tan suo* [An exploration of China's fiscal system], Shanghai: Shanghai renmin chubanshe, 1994
—— *Zhongguo hongguan caizheng zhengce yanjiu* [China's macro fiscal policy], Shanghai: Shanghai caijing daxue chubanshe, 1998
CPCCC 1998, *Zhonggong zhongyang guanyu nongye he nongcun gongzuo ruogan zhongda wenti de jueding* [Decision regarding several major problems in agricultural and rural work]
Diamond, L. J., *Developing Democracy: toward consolidation*, Baltimore MD: Johns Hopkins University Press, 1999
Ding, D., 'Wo guo xiang (zhen) cun fuzhai wenti de chengyin ji qi huajie: Anhui Sheng Suzhou Shi xiang (zhen) cun fuzhai qingkuang de diaocha' [Causes of and solutions to the problem of township- and village-level debt in China: an analysis of the debt situation in the townships, towns and villages of Suzhou City, Anhui Province], *Jingji Yanjiu Cankao* 72 (1744), 2003, pp. 39–44
Du, D. and Liu, J., 'He Kaiyin xishuo "er ci da baogan"' [He Kaiyin elaborates on the 'Second Household Responsibility System'], *Jingji luntan* 101 (20), 2001
Du, G., 'Jigou gaige chedi, bixu dangzheng fenkai' [For thorough organizational reform, Party and government must be separated], *Jiancha Ribao*, January 16, 2003. Online. Available HTTP: www.jcrb.com/zyw/n6/ca12185.htm (accessed on November 2, 2009)
Ebel, R. D. and Yilmaz, S., *Concept of Fiscal Decentralization and Worldwide Overview*, Washington DC: World Bank, 2002
Edin, M., 'State capacity and local agent control in China: CCP cadre management from a township perspective', *China Quarterly* 173, 2003, pp. 35–52
Fan, L., 'Anhui: cong shuifei gaige dao nongcun zonghe gaige' [Anhui: from the Tax and Fee Reform to comprehensive village reforms], *Sinanet*, March 6, 2005. Online. Available HTTP: http://finance.sina.com.cn/g/20050306/17591407141.shtml (accessed on November 1, 2009)
F County, RTFR Office, *F xian nongcun shuifei gaige shishi fang'an* [Programme for the implementation of the Rural Tax and Fee Reform in F County], May 5, 2000, internal reference
—— *F xian nongcun shuifei gaige juti caozuo banfa* [Concrete operating measures for the Rural Tax and Fee Reform in F County], February 11, 2001, internal reference
Fewsmith, J., 'Elite politics', in M. Goldman and R. MacFarquhar (eds), *The Paradox of China's post-Mao Reforms*, Cambridge MA: Harvard University Press, 1999, pp. 47–75
—— *China since Tiananmen: the politics of transition*, Cambridge and New York: Cambridge University Press, 2001
Friedman, E., Pickowicz, P. and Selden, M., *Chinese Village, Socialist State*, New Haven CT: Yale University Press, 1991
Fuhr, H., 'Institutional change and new incentive structures for development: can decentralization and better local government help?', *Welttrends* 25 (4), 2000, pp. 21–51
Gang, F., 'Market-oriented economic reform and growth of off-budget local public finance', in D. J. S. Brean (ed.), *Taxation in Modern China*, New York: Routledge, 1998, pp. 209–28

Gao, X., 'Nongcun shuifei gaige: redian wenti leng sikao' [The Rural Tax and Fee Reform: cool thoughts on a heated problem], in Z. He (ed.), *Jiceng minzhu he difang zhili chuangxin* [Grassroots democracy and innovations in local governance], Beijing: Zhongguo bianyi chubanshe, 2004, pp. 180–94

Gao, X. and Yang, Y., 'Cong nongcun shuifei gaige kan xiangzhen zhengfu gonggong chanpin de gongji' [Looking at the influence of the Rural Tax and Fee Reforms on the supply of public goods by township and town governments], *Gaige neican* 2003 (18), 2003, pp. 36–8

Göbel, C., 'Measuring and explaining the distributive effects of the Rural Tax and Fee Reform in Anhui Province, China', *China Perspectives* (2008/02), 2008, pp. 58–79

Guo, F., *Nongcun shuifu yu nongmin fudan* [The agricultural tax and the peasant burden], Beijing: Jingji ribao chubanshe, 2003

Guo, Z. and Bernstein, T. P., 'The impact of elections on the village structure of power: the relations between the village committees and the party branches', *Journal of Contemporary China* 13 (39), 2004, pp. 257–75

Guobanfa [1992] No. 42, *Guowuyuan bangongting guanyu jin yi bu zuo hao nongmin chengdan feiyong he laowu jiandu guanli gongzuo de tongzhi* [Notification by the general office of the State Council regarding the further improvement of the management and supervision of peasant fiscal burden and compulsory labour]

—— [2002] No. 25, *Guowuyuan bangonting guanyu zuohao 2002 nian guangda nongcun shuifei gaige shidian gongzuo de tongzhi* [Notification by the general office of the State Council concerning the orderly expansion of the RTFR pilot areas in 2002]

—— [2004] No. 68, *Guowuyuan bangongting guanyu chengli shenhua nongcun shuifei gaige shidian gongzuo zhidaoxing wenjian qicao xiaozu he zhuanti xiaozu de tongzhi* [Notice by the general office of the State Council regarding the establishment of authoring and expert groups for the drafting of guidance documents relating to the intensivization of RTFR pilot work]

Guobanhan [2006] No. 63, *Guowuyuan bangongting guanyu tongyi guowuyuan nongcun shuifei gaige gongzuo xiaozu ji qi bangongshi bianming de fuhan* [Affirmative reply by the general office of the State Council to the State Council RTFR working group and its office regarding its change of name]

Guofa [1993] No. 85, *Guowuyuan guanyu shixing fenshuizhi caizheng guanli tizhi de jueding* [Decision by the State Council regarding the introduction of a tax separation budget management system]

—— [1996] No. 12, *Guowuyuan pizhuan caizhengbu, guojia jihua weiyuanhui, shenjichu, Zhongguo renmin yinhang, jianchabu guanyu qingli jiancha yusuan wai zijin yijian de tongzhi* [Notification by the State Council concerning the commenting and transmission of the views of the Ministry of Finance, the State Planning Commission, the Audit Office, the People's Bank of China and the Ministry of Supervision regarding the cleaning up and investigation of extra-budgetary revenue]

—— [1996] No. 29, *Guowuyuan guanyu sjiaqiang yusuan wai zijin guanli de jueding* [Decision by the State Council regarding the strengthening of the management of extra-budgetary finances]

—— [2004] No. 21, *Guowuyuan guanyu zuohao 2004 nian shenhua nongcun shuifei gaige shidian gongzuo de tongzhi* [Notice by the State Council regarding doing well the intensivization of RTFR pilot work in 2004]

—— [2006] No. 34, *Guowuyuan guanyu zuohao nongcun zonghe gaige gongzuo youguan wenti de tongzhi* [Notice by the State Council regarding several problems relating to handling well the work in the rural comprehensive reform]

Guononggai [2002] No. 9, *Guowuyuan nongcun shuifei gaige gongzuo xiaozu, Guowuyuan jiuzheng hangye buzheng zhi feng bangongshi guanyu qieshi jiaqiang nongcun shuifei gaige shidian gongzuo de jinji tongzhi* [Urgent notification by the State Council's Rural Tax and Fee Reform working group and the government ethics office concerning realistic strengthening of work in the Rural Tax and Fee Reform pilots]

—— [2002] No. 10, *Guowuyuan nongcun shuifei gaige gongzuo xiaozu guanyu nongcun shuifei gaige shidian diqu zhanting qingshou nongmin shuifei weiqian gongzuo youguan wenti de jinji tongzhi* [Urgent notification by the State Council's Rural Tax and Fee Reform work group concerning questions regarding the temporary cessation of settling small tax and fee arrears with farmers in the Rural Tax and Fee Reform pilot areas]

Guononggaiban [2002] No. 30, *Guowuyuan nongcun shuifei gaige xiaozu bangongshi guanyu yinfa 'Xiang Huaicheng tongzhi zai quanguo nongcun shuifeigaige shidian gongzuo zuotanhui shang de jiangfa' he 'Du Qingling tongzhi zai quanguo nongcun shuifeigaige shidian gongzuo zuotanhui jieshu shi de jianghua'* [Notification by the office of the State Council's Rural Tax and Fee Reform work group's printing and distribution of 'Comrade Xiang Huaicheng's speech before the national symposium on work in the Rural Tax and Fee Reform pilot areas' and 'Comrade Du Qinglin's closing remarks at the national symposium on work in the Rural Tax and Fee Reform pilot areas']

Guowuling [1991] No. 92, *Nongmin chengdan feiyong he laowu guanli tiaoli* [Regulation for the management of peasant fiscal burden and compulsory labour]

Guowuyuan 1990, *Guowuyuan guanyu qieshi jianqing nongmin fudan de tongzhi* [Notification by the State Council concerning the realistic reduction of peasant burden]

Han, L., 'A commentary on "one, three, five" at mid-year', *Beijing Xinhua Domestic Service*, July 26, 1998, FBIS-CHI-98-210

Hauff, D., *Finanzausgleichspolitik in der Volksrepublik China. Eine empirische Analyse der Haushaltsreform von 1994 im Spannungsfeld zwischen Zentrale und Peripherie* [The politics of budget balancing in the People's Republic of China: an empirical analysis of the 1994 budgetary reform between the opposite poles of the centre and the periphery], Frankfurt am Main etc.: Peter Lange, 2003

He, G., *Zhongguo nongcun jinrong fazhan yu zhidu bianqian* [The development of rural credit in China and institutional change], Beijing: Zhongguo caizheng jingji chubanshe, 2005

He, K. and Sun, L. (eds.), *Zhongguo nongcun shuifei gaige chutan*, Beijing: Zhongguo Zhigong Chubanshe, 2000

He Xian zhengfu, *He Xian xiangzhen caizheng guanli fangshi gaige gongzuo qingkuang huibao* [Work status report for the township-level fiscal administration reform in He County], April 23, 2004, internal reference, 2004

Heberer, T., *Korruption in China. Analyse eines politischen, ökonomischen und sozialen Problems*, Opladen: Westdeutscher Verlag, 1991

—— 'Das politische System der VR China im Prozess des Wandels', in C. Derichs and T. Heberer (eds), *Einführung in die politischen Systeme Ostasiens*, Opladen: Leske & Budrich, 2003, pp. 19–121

—— *Private Entrepreneurs in China and Vietnam: social and political functioning of strategic groups*, Leiden and Boston MA: Brill, 2003

—— 'Korruption in China', in Alemann, U. (ed.) *Dimensionen politischer Korruption: Beiträge zum Stand der internationalen Forschung*, Wiesbaden: Verlag für Sozialwissenschaften, 2005, pp. 328–49

—— *Doing Business in Rural China: Liangshan's new ethnic entrepreneurs*, Seattle WA: University of Washington Press, 2007

—— 'Das politische System der VR China im Prozess des Wandels', in C. Derichs and T. Heberer (eds), *Einführung in die politischen Systeme Ostasiens*, Wiesbaden: VS Verlag, 2008

Heberer, T. and Taubmann, W., *Chinas ländliche Gesellschaft im Umbruch. Urbanisierung und sozioökonomischer Wandel auf dem Lande*, Opladen and Wiesbaden: Westdeutscher Verlag, 1998

Heerink, N., Kuiper, M. and Shi, X., 'China's new rural income support policy: impacts on grain production and rural income inequality', *China and World Economy* 14 (6), 2006, pp. 58–69

Heilmann, S., 'From local experiments to national policy: the origins of China's distinctive policy process', *China Journal* 59, 2008, pp. 1–30

—— 'Policy experimentation in China's economic rise', *Studies in Comparative International Development* 43 (1), 2008, pp. 1–26

Ho, S. P. S., *Rural China in Transition: non-agricultural development in rural Jiangsu, 1978–1990*, Oxford and New York: Oxford University Press, 1994

Hood, C., *The Art of the State: culture, rhetoric, and public management*, Oxford and New York: Clarendon Press, 1998

Hsiao, K. H. Y. H., *The Government Budget and Fiscal Policy in Mainland China*, Seattle WA: University of Washington Press, 1987

Hutchcroft, P., 'Centralization and decentralization in administration and politics: assessing territorial dimensions of authority and power', *Governance: an International Journal of Policy and Administration* 14 (1), 2001, pp. 23–53

Jia, K. Z., Quanhou, 'Jiefu zhihou: nongcun shuifei gaige dai jiejue de wenti ji duice tantao' [After burden reduction: problems of the Rural Tax and Fee Reform waiting to be solved and an exploration of solutions], *Caizheng Yanjiu* (5), 2002, pp. 5–8

Jiang, Zemin, *Jiang Zemin zhuxi 1998 nian 9 yue 25 ri zai Anhui kaocha gongzuo shi de jiang hua* [[Party] chairman Jiang Zemin's speech during an investigative visit to Anhui on September 25, 1998], September 25, 1998, internal reference, 1998

Jiaojian [2001] No. 3, *Anhui sheng jiaoyuting, Anhui sheng jianchating, Zhongguo baoxian jiandu guanli weiyuanhui Hefei chipaiyuan banshichu guanyu yanjin nongcun zhongxiao xuexiao daiban xuesheng baoxian de tongzhi* [Notification by the Ministry of Education, the Procuratorial Department of Anhui province, and the Hefei representative of the Committee for Administration and Supervision for China's Insurance regarding the strict illegality of rural primary and middle schools handling student insurance]

Kelliher, D., 'The Chinese debate over village self-government', *China Journal* 37, 1997, pp. 63–86

Kenis, P. and Schneider, V., 'Verteilte Kontrolle: Institutionelle Steuerung in modernen Gesellschaften', in P. Kenis and V. Schneider (eds), *Organisation und Netzwerk. Institutionelle Steuerung in Wirtschaft und Politik*, Frankfurt and New York: Campus Verlag, 1996, pp. 9–43

Kennedy, J. J., 'From the Tax-for-fee reform to the abolition of agricultural taxes: the impact on township governments in north-west China', *China Quarterly* 189, 2007, pp. 43–59

Kipnis, A., 'Within and against peasantness: backwardness and filiality in rural China', *Comparative Studies in Society and History* 37 (1), 1995, pp. 110–35

Lam, T.-C. and Perry, J. L., 'Service organizations in China: reform and its limits', in P. N.-S. Lee and C. W.-H. Lo (eds), *Remaking China's Public Management*, Westport CT: Quorum Books, 2001, pp. 19–40

Lam, W., 'Hu's new deal and the third plenary session of CCP's seventeenth central committee', *China Brief* 8 (20), 2008. Online. Available HTTP: http://www.jamestown.org/programs/chinabrief/single/?tx_ttnews[tt_news] = 5230&tx_ttnews[backPid] = 168&no_cache = 1 (accessed on October 19, 2009)

Lan, Z., 'Local government reform in the People's Republic of China: stipulations, impact, cases and assessment', *Chinese Public Administration Review* 1 (3–4), 2002, pp. 209–20

Lardy, N., 'China's worsening debts', *Financial Times*, June 22, 2001. Online. Available HTTP: http://www.brookings.edu/opinions/2001/0622china_lardy.aspx (accessed on November 2, 2007)

Lauth, H.-J., *Im Schatten demokratischer Legitimität. Informelle Institutionen und politische Partizipation im interkulturellen Demokratienvergleich*, Opladen: Westdeutscher Verlag, 1999

—— 'Informal institutions and democracy', *Democratization* 7 (4), 2000, pp. 21–50

Li, C., *Wo xiang zongli shuo shihua* [I spoke the truth to the Premier], Beijing: Guangming ribao chubanshe, 2002

—— 'Nongmin de pinkun: tudi, quanli, ziyou fazhan' [Peasant poverty: land, power, development of freedom], in Y. Zhou (ed.), *Nongmin! Nongmin!*, Guangzhou: Huacheng chubanshe, 2004

—— 'Think national, blame local: central–provincial dynamics in the Hu era', *China Leadership Monitor* (17), 2006, pp. 1–24. Online. Available HTTP: http://media.hoover.org/documents/clm17_lc.pdf, (accessed on October 30, 2009)

Li, F., 'Zhongguo nongmin de zhengzhi xuanze' [Political choices for China's farmers], manuscript, 2004

Li, H., 'Finance in rural China', in Rural Development Institute of the Chinese Academy of Social Sciences (ed.), *Clashes and Balances: the transition of the economy and society in rural China*, Beijing: Foreign Languages Press, 2006, pp. 94–120

Li, J., *Taxation in the People's Republic of China*, New York: Praeger, 1991

Li, L., 'Political trust in rural China', *Modern China* 30 (2), 2004, pp. 228–58

Li, L. C., 'Provincial discretion and national power: investment policy in Guangdong and Shanghai, 1978–93', *China Quarterly* 152, 1997, pp. 778–804

—— 'Differentiated actors: central and local actors in China's rural tax reforms', *Modern Asian Studies* 40 (1), 2006, pp. 151–74

—— 'Working for the peasants? Strategic interactions and unintended consequences in the Chinese rural tax reform', *China Journal* 57, 2007, pp. 89–106

Li, Y., 'Hubei sheng xiangzhen zonghe peitao gaige dui nongcun hexie shehui goujian de jiji yingxiang' [The positive influence of Hubei province's township and town comprehensive complementary reforms on building a harmonious society in the countryside], *Guojia Xingzheng Xueyuan*. Online. Available HTTP: www.nsa.gov.cn/download/liyoucai.doc (accessed on October 23, 2005)

Lieberthal, K., *Governing China: from revolution through reform*, New York: W. W. Norton, 1995

Lieberthal, K. and Lampton, D. M., *Bureaucracy, Politics, and Decision-making in post-Mao China*, Berkeley CA: University of California Press, 1992

Liu, S., 'Compulsory education must on no account be weakened: sidelights on Premier Zhu Rongji's inspection of compulsory education in the Anhui rural areas', Beijing Xinhua Domestic Service, July 22, 2001, FBIS-CHI-2001-0722

Liu, Z., 'Zhongguo zhengfu jigou cunzai de zhuyao wenti' [The main problems in China's government structure], *Zhanlüe yu guanli* 5 (36), 1999, pp. 90–5

—— *Zhongguo difang shuizhi gaijian* [Local taxation in China], Beijing: Zhongguo jingji chubanshe, 2005

Lo, C. W.-H., Lo, J. M.-K. and Cheung, K.-C., 'Service organizations in the environmental governance system of the People's Republic of China', in P. N.-S. Lee and C. W.-H. Lo (eds), *Remaking China's Public Management*, Westport CT: Quorum Books, 2001, pp. 41–63

Lu, Q., 'Chen Xiwen: xin nongcun jianshe huiji yi wan nongmin. Zhongyang gei si xiang buzhu' [Chen Xiwen: the construction of the New Socialist Countryside benefits hundreds of millions of peasants; the centre gives four kinds of subsidies], *Renmin Ribao Haiwai ban*, May 22, 2007. Online. Available HTTP: http://www.gov.cn/jrzg/2007-05/22/content_621803.htm (accessed on November 1, 2009)

Lü, X., 'The politics of peasant burden in reform China', *Journal of Peasant Studies* 25 (1), 1997, pp. 113–38

—— *Cadres and Corruption: the organizational involution of the Chinese Communist Party*, Stanford CA: Stanford University Press, 2000

Luard, T., 'China village democracy skin deep', BBC Online, October 10, 2005. Online. Available HTTP: http://newsvote.bbc.co.uk/mpapps/pagetools/print/news.bbc.co.uk/2/hi/asia-pacific/4319954.stm (accessed on October 10, 2009)

Ma, X., Huang, H. and Jiang, C., *Wo guo nongcun shuifei gaige yanjiu* [The Rural Tax and Fee Reform in China], Beijing: Zhongguo jihua chubanshe, 2002

MacIntyre, A., Khan, M. H. and Jomo, K. S., 'Funny money: fiscal policy, rent-seeking and economic performance in indonesia', in M. H. Khan and K. S. Jomo (eds), *Rents, Rent-seeking and Economic Development: theory and evidence in Asia*, Cambridge etc.: Cambridge University Press, 2000, pp. 248–73

Merkel, W., *Systemtransformation. Eine Einführung in die Theorie und Empirie der Transformationsforschung* [System transformation: an introduction to the theory and practice of transformation research], Opladen: Leske & Budrich, 1999

M Town, Finance Office, *2006 nian M zhen caizheng shouzhi* [Budgetary revenues and expenditures of M Town for 2006], internal reference, 2006

National Bureau of Statistics, *China Statistical Yearbook*, Beijing: Zhongguo tongji chubanshe, various years

Nohlen, D., *Wahlrecht und Parteiensystem. Zur Theorie der Wahlsysteme*, Opladen: Leske & Budrich, 2003

O'Brien, K. and Li, L., 'Popular contention and its impact in rural China', *Comparative Political Studies* 38 (3), 2005, pp. 235–59

Oi, J., *Rural China Takes Off: institutional foundations of economic reform*, Berkeley CA: University of California Press, 1999

Oi, J. and Zhao, S., 'Fiscal crisis in China's townships: causes and consequences', in E. J. Perry and M. Goldman (eds), *Grassroots Political Reform in Contemporary China*, Cambridge MA: Harvard University Press, 2007, pp. 75–96

Oi, J. C., *Rural China Takes Off: institutional foundations of economic reform*, Berkeley CA: University of California Press, 1999

Oi, J. C. and Walder, A. G., *Property Rights and Economic Reform in China*, Stanford CA: Stanford University Press, 1999

Pan, P. P., 'Chinese peasants attacked in land dispute: at least six die as armed thugs assault villagers opposed to seizure of property', *Washington Post*, June 15, 2005, p. A12

Powell, S., *Agricultural Reform in China: from communes to commodity economy, 1978–1990*, Manchester and New York: Manchester University Press, 1992

Pu, X., *Zhonghua renmin gongheguo zhengzhi tixi* [The political system of the People's Republic of China], Shanghai: Shanghai renmin chubanshe, 1999

Qu, Z., Li, X. and Wang, X., *Zhongguo shehui zhuyi xin nongcun jianshe yanjiu* [New rural construction in socialist China], Beijing: Shehui kexue wenxian chubanshe, 2006

Riskin, C., Zhao, R. and Li, S., *China's Retreat from Equality: income distribution and economic transition*, Armonk NY: M. E. Sharpe, 2001

Sartori, G., 'Concept misformation in comparative politics', *American Political Science Review* 64 (4), 1970, pp. 1033–53

Schubert, G., 'Village Elections in the PRC: A Trojan Horse of Democracy?', *Duisburg Working Paper on East Asian Studies* 2002 (19), 2002

Shambaugh, D. L., *Modernizing China's military: progress, problems, and prospects*, Berkeley CA: University of California Press, 2002

Shengjiuban [2002] No. 1, *Anhui Sheng renmin zhengfu jiuzheng hangye buzheng zhi feng bangongshi 'guanyu chengli Sheng she nong anjian chachu bangongshi de tongzhi'* [Notification by the Bureau of Government Ethics of Anhui Province 'regarding the establishment of an office for the investigation and treatment of rural cases occurring in Anhui province']

Shi, W., *Gong xuan yu zhi xuan: xiang zhen ren da xuan ju zhi du yan jiu*, Beijing: Zhongguo she hui ke xue chu ban she, 2000

Shue, V., 'State power and social organization in China', in J. S. Migdal, A. Kohli and V. Shue (eds), *State Power and Social Forces: domination and transformation in the Third World*, Cambridge and New York: Cambridge University Press, 1994, pp. 65–88

Solinger, D., 'The Creation of a New Underclass in China and its Implications', *Centre for the Study of Democracy* Paper 05 (10), 2005

State Council General Office, *Report on the Work of the Government*, 2005. Online. Available HTTP: http://www.gov.cn/english/official/2005-7/29/content_18351.htm (accessed on June 23, 2009)

Taihe County, RTFR Office, 'Anhui Taihe xian nongcun shuifei gaige shijian' [The experience of the RTFR in Taihe County, Anhui Province], internal document, 1994

Tang, W., *Public Opinion and Political Change in China*, Stanford CA: Stanford University Press, 2005

Tanigaki, S., 'Japanese Economy Revived: Challenges and the Way Ahead: fiscal consolidation is the biggest remaining challenge', paper presented at the Japan Society, New York, January 9, 2006

Taube, M., *Institutionelle Lebenszyklen im chinesischen Transformationsprozess. Aufstieg und Fall der Township Village Enterprises*, Duisburg: Diskussionsbeiträge der Mercator School of Management der Universität Duisburg-Essen 322, 2007

Treisman, D., 'Political decentralization and economic reform: a game-theoretic analysis', *American Journal of Political Science* 43 (2), 1999, pp. 488–517

Usher, D., 'The dynastic cycle and the stationary state', *American Economic Review* 79 (5), 1989, pp. 1031–44

Wanbanfa [2000] No. 11, *Zhonggong Anhui shengwei bangongting, Anhui sheng renmin zhengfu bangongting guanyu yinfa Anhui sheng nongcun shuifei gaige youguan peitao wenjian de tongzhi* [Notice of the Communist Party Anhui Provincial Committee's general office and the Anhui Provincial government's general office concerning the printing and distribution of supplementary documents relating to the RTFR in Anhui province]

—— [2000] No. 14, *Zhonggong Anhui sheng bangongting, Anhui sheng renmin zhengfu bangongting guanyu zhuanfa 'sheng wujia ju, sheng caizheng ting, sheng jianchating, sheng nongwei, sheng zhengfu jiufengban guanyu qingli zhengdun she nong shoufei, qieshi jianqing nongmin fudan de yijian' de tongzhi* [General Office of the Communist Party Anhui Provincial Committee, general office of the Anhui Provincial Government, 'Views of the Provincial Pricing Department, Finance Department, Procurate, Farmers' Comittee and the Ethics Office regarding the clearance and rectification of agricultural fees and the thorough lessening of the farmers' burden']

—— [2000] No. 21, *Zhonggong Anhui shengwei bangongting, Anhui sheng renmin zhengfu bangongting 'guanyu renzhen zuohao wu baohu shenghuo anpai, shoumie diqu nongyeshui jianmian he qieshi jiaqiang nongcun shui, dian deng shengchan jingyingxing feiyong shouqu guanli gongzuo de jinji tongzhi'* [Office of the Communist Party Anhui Provincial Committee and the Office of the Anhui Provincial Government's 'Urgent notification regarding the dilligent improvement of the five "to be protecteds'" living conditions, agricultural tax reductions in areas stricken by disaster and the thorough strengthening of rural water, electricity and other supplies as well as service fees collection and administration work']

—— [2000] No. 41, *Zhonggong Anhui sheng bangongting 'guanyu zai nongcun shuifei gaige shidian guocheng zhong qieshi zuo hao jianqing nongmin fudan gongzuo de tongzhi'* [General office of the Communist Party Anhui Provincial Committee, 'Notification concerning the thorough improvement of burden reduction work in the RTFR pilots']

—— [2001] No. 7, *Zhonggong Anhui shengwei bangongting, Anhui sheng renmin zhengfu bangongting guanyu yinfa 'guanyu jin yi bu jiaqiang he guifan nongye shuishou zhengshou guanli gongzuo de yijian' deng nongcun shuifei gaige youguan peitao wenjian de tongzhi* [Notice of the Communist Party Anhui Provincial Committee's general office and the Anhui Provincial government's general office concerning the printing and distribution of the 'Views regarding the further strengthening and standardization of agricultural tax collection and administration work' and other supplementary documents relating to the RTFR]

—— [2001] No. 18, *Zhonggong Anhui shengwei bangongting, Anhui sheng renmin zhengfu bangongting 'guanyu jiang nongcun chuzhong, xiaoxue laoshi gongzi shangshou dao xian guanli de tongzhi'* [General office of the Anhui Provincial Communist Party committee and the general office of the Anhui provincial government, 'Notification regarding the transfer of the administration of rural middle and primary schoolteachers' salaries to the county level']

—— [2001] No. 19, *Zhonggong Anhui shengwei bangongting, Anhui sheng renmin zhengfu bangongting guanyu yinfa 'guanyu fangzhi xiang cun liang ji fasheng xin de buliang*

zhaiwu de yijian' deng nongcun shuifei gaige youguan peitao wenjian de tongzhi [Notice of the Communist Party Anhui provincial committee's general office and the Anhui provincial government's general office concerning the printing and distribution of the 'Views regarding the prevention of new bad credits at the township and village levels' and other supplementary documents related to the RTFR]
—— [2002] No. 10, *Anhui Sheng renmin zhengfu bangongting zhuanfa Guowuyuan bangongting zhuanfa Nongyebu deng bumen guanyu 2002 nian jianqing nongmin fudan gongzuo yijian de tongzhi* [Notification by the general office of the government of Anhui province regarding the distribution of the notification by the office of the State Council regarding the distribution of the views of the Ministry of Agriculture and other Ministries concerning peasant burden reduction in 2002]
Wanbanmingdian [2000] No. 47, *Zhonggong Anhui sheng bangongting, Anhui sheng renmin zhengfu bangongting 'guanyu renzhen zuo hao nongyeshuishou zhengshou guanli gongzuo' de tongzhi* [General office of the Communist Party Anhui Provincial Committee, general office of the Anhui provincial government, 'Notification regarding the diligent implementation of agricultural tax collection and administrative work']
—— [2001] No. 96, *Zhonggong Anhui shengwei, Anhui sheng renmin zhengfu guanyu zhuanfa 'sheng jiwei, shengwei xuanchuanbu, sheng jianchating, sheng xinwenchubanju guanyu zhengdun baokan faxing chixu, qieshi jianqing cunji fudan de ruogan yijian' de tongzhi* [Notice of the Communist Party Anhui Provincial Committee and the Anhui Provincial Government regarding distribution of 'Several opinions of the Provincial Disciplinary Committee, Propaganda Department, Procurate and Publication Department regarding the rectification of the mechanisms for the distribution of publications and the thorough reduction of the village-level burden']
Wandanglianfa [2001] No. 6, *Anhui sheng dang'anju, Anhui sheng nongcun shuifei gaige lingdao xiaozu bangongshi guanyu yinfa 'Anhui sheng nongcun shuifei gaige dang'an guanli banfa' de tongzhi* [Notification by the Bureau of Files and the office of the RTFR LSG regarding the printing and distribution of 'measures regarding the administration of files related to the RTFR in Anhui province']
Wanfa [2000] No. 10, *Zhonggong Anhui shengwei, Anhui sheng renmin zhengfu guanyu zai quan sheng fanwei ne kaizhan nongcun shuifei gaige shidian gongzuo de tongzhi* [Notice of the Communist Party Anhui provincial committee and the Anhui provincial government regarding the beginning of RTFR pilot work in the whole province]
—— [2000] No. 15, *Zhonggong Anhui shengwei, Anhui sheng renmin zhengfu guanyu xiangzhen dang zheng jigou gaige de shishi yijan* [Views of Communist Party Anhui provincial committee and Anhui provincial government regarding the implementation of party and government structural reforms at the township level]
—— [2000] No. 16, *Zhonggong Anhui shengwei, Anhui sheng renmin zhengfu guanyu xiangzhen shiye danwei jigou gaige de yijian* [Views of the Communist Party Anhui Provincial committee and the Anhui Provincial government regarding the implementation of the reform of Party and government structures at the township and town levels]
—— [2000] No. 22, *Zhonggong Anhui sheng bangongting, Anhui sheng renmin zhengfu bangongting 'guanyu yinfa "guanyu jin yi bu tiaozheng nongcun zhongxiaoxue buju de yijian" deng nongcun shuifei gaige youguan peitao wenjian de tongzhi'* [General office

of the Communist Party Anhui provincial committee, general office of the Anhui provincial government, 'Notification regarding the printing and distribution of the 'Views regarding the further rectification of village primary school arrangements' and other supplementary documents concerning the RTFR']

—— [2001] No. 6, *Zhonggong Anhui shengwei, Anhui sheng renmin zhengfu guanyu zuo hao 2001 nian quan sheng nongcun shuifei gaige gongzuo de yijiani* [View of the Communist Party Anhui provincial committee and the Anhui Provincial government regarding the well handling of RTFR work in the whole province]

—— [2002] No. 3, *Zhonggong Anhui shengwei, Anhui sheng renmin zhengfu guanyu zuo hao 2002 nian quan sheng nongcun shuifei gaige gongzuo de yijian* [Views of the Communist Party Anhui Provincial committee and the Anhui Provincial government regarding doing well the province-wide RTFR work in 2002]

—— [2004] No. 9, *Anhui sheng renmin zhengfu guanyu quanmian quxiao nongye shui fujia de tongzhi* [Notice by the government of Anhui Province regarding the complete abolition of the supplement to the Agricultrural Tax]

Wang, B. and Zhao, Y., 'Nongcun shuifei gaige dui zhongxi bu xiangzhen caili yingxiang de shijian yanjiu – jiyu 4 sheng 8 xian chouyang diaocha shuju de fenxi' [Empirical research on the fiscal effects of the RTFR on central and western Chinese townships and towns], *Guanli Shijie* 2006 (11), 2006

Wang, S., 'China's 1994 fiscal reform: an initial assessment', *Asian Survey* 37 (9), 1997, pp. 801–17

Wang, S. and Hu, A., *Zhongguo guojia nengli baogao* [Report on China's state capacity], Shenyang: Liaoning renmin chubanshe, 1993

Wang, T., 'Wo guo caizheng tizhi gaige ying jieshou gonggong caizheng lilun' [The reform of China's fiscal system must conform to theories of public finance], *Neibu canyu* 47 (740), 2004, pp. 2–8

Wannongjian [2001] No. 2, *Anhui sheng nongmin fudan jiandu guanli lingdao xiaozu 'guanyu jin yi bu guanfan he wanshan nongmin fudan jianduka zhidu de tongzhi'* [Notification by the leading small group for peasant burden supervision and administration of Anhui province regarding the further standardization and fine-tuning of the peasant burden supervision card system]

—— [2001] No. 3, *Anhui sheng nongmin fudan jiandu guanli lingdao xiaozu 'guanyu queding jianqing nongmin fudan zhongdian gongzuo zeren danwei de tongzhi'* [Notification by the leading small group for peasant burden supervision and administration of Anhui province regarding the confirmation of the units with special responsibilities in peasant burden reduction]

Wanting [2001] No. 17, *Zhonggong Anhui shengwei bangongting, Anhui sheng renmin zhengfu bangongting guanyu yinfa 'guanyu tiaozheng nonge techan shui youguan zhengce de yijian' de tongzhi* [Notice of the Communist Party Anhui provincial committee's general office and the Anhui provincial government's general office concerning the printing and distribution of the 'Views regarding the policies concerning the adjustment of the agricultural special products tax']

Wanzheng [2001] No. 99, *Anhui sheng renmin zhengfu zhuanfa Sheng wujiaju deng liu bumen guanyu zhuanxiang zhili nongmin jianfang shoufei qieshi jianqing nongmin fudan yijian de tongzhi* [Notification by the government of Anhui Province regarding the distribution of the views of the provincial pricing and six other departments regarding the special handling of peasant house-building fees and the realistic reduction of the peasant burden]

Wanzhengban [2001] No. 60, *Anhui sheng renmin zhengfu bangongting guanyu zhuanfa 'guowuyuan bangongting guanyu qieshi zu hao dangqian qianjing nongmin fudan gongzuo de tongzhi'* [Notification by the general office of the government of Anhui Province regarding the distribution of the 'Notification by the general office of the State Council concerning the thorough implementation of the present peasant burden reduction work']

—— [2001] No. 79, *Anhui sheng renmin zhengfu bangongting guanyu zhuanfa Sheng nongye weiyuanhui guanyu jin yi bu wending nongcun tudi chengbao guanxi wanshan nongye chengbao hetong guanli ruogan yijian de tongzhi* [Distribution by the general office of the government of Anhui Province of the notification by the provincial agricultural committee regarding several opinions concerning the further stabilization of village land contractual relationships and perfecting the administration of the household responsibility system]

—— [2001] No. 84, *Anhui sheng renmin zhengfu bangongting zhuanfa Sheng wujiaju deng wu bumen guanyu jiaqiang nongcun hunying dengji, jihua shengyu deng shoufei guanli yijian de tongzhi* [Notification by the general office of the government of Anhui Province regarding the distribution of the views of the provincial pricing and five other departments regarding the improved management of the marriage registration fee and the birth control fee]

—— [2002] No. 5, *Anhui sheng renmin zhengfu bangongting guanyu jiji wentuo de huajie cunji zhaiwu de yijian* [Views of the general office of the government of Anhui Province regarding the proactive and stable dissolution of village-level debts]

Wanzhengmingdian [2000] No. 35, *Anhui sheng renmin zhengfu 'guanyu jinyibu guifan nongye shui, nongye techan shui ji qi fujia zhengguan gongzou de jinji tongzhi'* [Anhui provincial government's 'Urgent notification regarding further standardization of the collection and administration of the agricultural tax and the agricultural special products tax and their supplementary charges']

—— [2001] No. 6, *Anhui sheng renmin zhengfu guanyu zhanting xiang nongmin shouqu Huabei pingyuan gaizao he Pishihang Chaohu guanggai xiangmu yiji nongye zonghe kaifa tudi zhili xiangmu huan zhai zijin de jinji tongzhi* [Urgent notification of Anhui provincial government concerning the temporary cessation of levying funds from the peasants to repay the loans for the Huaibei Plain Improvement, the Pishihang Chaohu area development project and the comprehensive rural development project]

—— [2001] No. 9, *Anhui sheng renmin zhengfu 'guanyu jiaqiang nongcun yiwu jiaoyu shoufei guanli jianjue zhizhi luan shoufei de tongzhi'* [Notification of Anhui provincial government concerning the strengthening of the administration of rural compulsory education fees and the permanent cessation of irregular fees]

—— [2001] No. 20, *Anhui Sheng renmin zhengfu guanyu jin yi bu luoshi nongcun shuifei gaige zhengce zuo hao wending nongmin fudan gongzuo de jinji tongzhi* [Urgent notification by the government of Anhui Province regarding the further implementation of the RTFR policies and the improved stabilization of peasant burdens]

Wen, Jiabao, *Wen Jiabao fuzongli zai Anhui guancha nongcun shuifei gaige shidian gongzuo shi de zhongyao jianghua zhaiyao* [Summary of Vice-premier Wen Jiabao's important speech at his inspection of the Rural Tax and Fee Reform pilot work in Anhui Province], April 13, 2000, internal reference, 2000

White, L. T., *Unstately Power*, Armonk NY: M. E. Sharpe, 1998

Whiting, S. H., *Power and Wealth in rural China: the political economy of institutional change*, New York and Melbourne: Cambridge University Press, 2001

Whiting, S. H., McMillan, J. and Naughton, B., 'Contract incentives and market discipline in China's rural industrial sector', in J. McMillan and B. Naughton (eds), *Reforming Asian Socialism: the growth of market institutions*, Ann Arbor MI: University of Michigan Press, 1996, pp. 63–110

Wong, C., *Financing Local Government in the People's Republic of China*, Hong Kong and New York: Oxford University Press, 1997

Wong, C. P. W. and Brean, D. J. S., 'Fiscal dualism in China: gradualist reform and the growth of off-budget finance', in D. J. S. Brean (ed.), *Taxation in Modern China*, New York and London: Routledge, 1998, pp. 187–208

Wong, C., Heady, C. J. and Woo, W. T., *Fiscal Management and Economic Reform in the People's Republic of China*, Hong Kong and New York: Oxford University Press, 1995

Work, R., 'Overview of Decentralization Worldwide: a stepping stone to improved governance and human development', paper presented at the second International Conference on Decentralisation ('Federalism: The Future of Decentralizing States?'), Manila, Philippines, July 25–27, 2002

World Bank, *China: National Development and Sub-national Finance: a review of provincial expenditures*, World Bank Report 22951-CHA, April 9, 2002,

—— *East Asia Decentralizes: making local government work*, Washington, DC: World Bank, 2005. Online. Available HTTP: http://www.loc.gov/catdir/toc/fy055/2005047796.html (accessed on December 1, 2007)

—— *World Development Indicators Database*, 2005. Online. Available HTTP: http://devdata.worldbank.org/data-query/ (accessed on December 1, 2007)

Wu, L., 'Nongcun shuifei gaige dui xiangzhen caizheng de yingxiang ji qi houguo: yi Anhui sheng wei li' [The effects of the Rural Tax and Fee Reform on township and town finances and their consequences, taking Anhui Province as an example]. Online. Available HTTP: http://www.usc.cuhk.edu.hk/wk_wzdetails.asp?id = 2114 (accessed on June 23, 2009)

Xia, Z., 'Nongcun shuifei gaige zhidu kuangjia jiben queli' [The basic establishment of the RTFR systemic framework], *Xinhua Kuaijiwang*. Online. Available from the author

Xiang, Huaicheng, *Xiang Huaicheng buzhang zai quanguo nongcun shuifei gaige gongzuo zuotanhui shang de jianghua zhaiyao* [Summary of Minister Xiang Huaicheng's speech at the symposium on the nationwide Rural Tax and Fee Reform work], November 4, 2000, internal reference, 2000

—— *Xiang Huaicheng buzhang zai quanguo nongcun shuifei gaige shidian gongzuo huiyi shang zongjie jianghua zhaiyao* [Summary of Minister Xiang Huaicheng's summarizing speech at the conference on the nationwide Rural Tax and Fee Reform Pilot work], February 19, 2001, internal reference, 2001

—— *Xiang Huaicheng tongzhi zai quanguo nongcun shuifei gaige shidian gongzuo zuotanhui shang de jianghua* [Comrade Xiang Huaicheng's speech at the national Rural Tax and Fee Reform Pilot work symposium], October 24, 2002, internal reference, 2002

Xiang, J., *Zouchu 'Huang Zongxi dinglü' de guaiquan: Zhongguo nongcun shuifei gaige de diaocha yu yanjiu* [Breaking out of the viscious cycle of Huang Zongxi's law: investigation and research on the RTFR in China], Xi'an: Xibei daxue chubanshe, 2004

Xiao, T. and Li, C., *Zhongguo xiangcun baogao: zhengfu xingwei yu xiangcun jianshe yanjiu* [Report on rural China: researching government behaviour and rural construction], Shanghai: Xu lin chubanshe, 2005

Xie, Q., *Xian zhengfu guanli: Wanning xian diaocha* [County government management: an investigation of Wanning County], Beijing: Guangbo dianshi chubanshe, 1994

Xu, F., 'Quxiao diji shi ru shiye, sheng guan xian jiang you shidian' [Abandoning prefectoral-level cities possible, pilot areas for provincial adminstration of counties set up], *Nanfang Zhoumo*, September 15, 2005

Yep, R., 'Can "tax for fee" reform reduce rural tension in China? The process, progress and limitations', *China Quarterly* 177, 2004, pp. 42–70

Yu, J., *Yuecun zhengzhi. Zhuanxingqi Zhongguo xiangcun zhengzhi jiegou de bianqian* [Politics in Yue Village: the changes in rural political structures in transforming China], Beijing: Shangwu yinshu guan, 2001

Yu, Q., 'Nongcun shuifei gaige de diaojiu ji xin jucuo' [Survey and new measures on the reform of taxes and fees in the countryside], *Huabei gongxueyuan xuebao (sheke ban)* 19 (3), 2003, pp. 84–86

Zhang, D., 'Wo kan nongcun di san ci geming' [My viewpoint on the Third Revolution in the countryside], *Shehui kexue zhanxian* 2003 (2), 2003, pp. 167–70

—— '"Bi shui" hai zai jixu' [Tax enforcement still continues], *Gaige neican* (31), 2003, pp. 20–4

—— 'Shuifei gaige hou de de "bishui xiaoying"' [The 'tax enforcement phenomenon' after the RTFR], in C. Li and L. Dong (eds), *Shuifei gaige beijing xia de xiangzhen tizhi yanjiu*, Wuhan: Hubei renmin chubanshe, 2004, pp. 326–30

Zhang, P., *Anhui nongcun shuifei gaige shijian yu tansuo* [Practice and exploration of Anhui's Rural Tax and Fee Reform], Beijing: Dangdai Chubanshe, 2001

Zhang, P. (ed.), *Anhui nongcun shuifei gaige shijian yu tansuo* (Beijing: Dangdai Chubanshe, 2001)

Zhao, S., 'Cunzhuang yu fudan-yi xie difang nongmin fudan guozhong zhuangkuang de fenxi' [Villages and burden: an analysis of several locations where the peasant burden is too heavy], *Guowuyuan fazhan yanjiu zhongxin xinxi wang*, October 31, 1999. Online. Available HTTP: http://www.drcnet.com.cn/DRCnet.Channel.Web/expert/showdoc.asp?doc_id = 137697 (accessed on October 17, 2005)

—— 'Xiangzhen gaige: jiantao yu zhanwang' [Township and town reforms: exploration and prospects], in X. Wu, X. Lu and P. Li (eds), *2005 nian: zhongguo shehui xingshi fenxi yu yuce* [Analysis and forecast of China's social development, 2005], Beijing: Shehui kexue wenxian chubanshe, 2005, pp. 65–78

Zhong, Y., *Local Government and Politics in China: challenges from below*, Armonk NY: M. E. Sharpe, 2003

Zhongbanfa [2000] No. 30, *Zhonggong zhongyang bangongting, guowuyuan bangongting guanyu shi xian xiang renyuan bianzhi jingjian de yijian* [Opinion of the general office of the Chinese Party's central committee and the general office of the State Council concerning the reduction of staff on the government payroll in cities, counties and townships]

Zhongfa [1990] No. 16, *Zhonggong zhongyang, guowuyuan guanyu jianjue zhizhi luan shoufei, luan fakuan he ge zhong tanpai de jueding* [Decision of the central committee of the CCP and the State Council concerning the resolute cessation of wanton fee-taking, wanton fining and all kinds of apportionments]

—— (1995) No. 4, *Zhonggong zhongyang guanyu yinfa 'Dang Zheng lingdao ganbu xuanbo rnyong gongzuo zhanxing tiali' de tongzhi* [Notification by the Central Committee of the CCP regarding the printing and distribution of the 'Provisional regulation for the selection and appointment of Party and state leading cadres']

—— [1996] No. 13, *Zhonggong zhongyang guowuyuan guanyu qieshi zuo hao nongmin fudan gongzuo de jueding* [Decision by the Central Committee of the CCP and the State Council to do good work in alleviating peasants' financial burden]

—— [2000] No. 7, *Zhonggong zhongyang, guowuyuan guanyu jinxing nongcun shuifei gaige shidian gongzuo de tongzhi* [Notification by the central committee of the Communist Party of China and the State Council concerning the advancement of work in RTFR pilot areas]

—— [2006] No. 1, *Zhonggong zhongyang guowuyuan guanyu jiji fazhan xiandai nongye zashi tuijin shehui zhiyi xin nongcun jianshe de ruogan yijian* [Several opinions of the central party committee of the Communist Party and the State Council regarding the proactive development of a modern agriculture and solidly promoting the construction of a new socialist countryside]

Zhongguo Shekeyuan Caimaosuo Ketizu, *Difang zhengfu zhaiwu: laiyuan, goucheng, yingxiang ji huajie duice* [Local government debt: origins, composition, influence and clearance counter-policies], Beijing: CASS, 2004

Zhonghua Renmin Gongheguo Guowuyuan ling 125, *Guojia Gongwuyuan zhanxing tiaoli* [Provisional regulation for government employees]

Zhou, T., *Zhongguo zhengzhi tizhi gaige* [Political system reform in China], Beijing: Zhongguo shuili shuidian chubanshe, 2004

Zhou, X., 'Bengkui de bianyuan: Zhongguo nongcun de jiceng zhengfu zhaiwu weiji' [On the brink of collapse: the debt crisis of grassroots governments in rural China], 2004. Online. Available HTTP: http://www.usc.cuhk.edu.hk/wk_wzdetails.asp?id = 3591 (accessed on October 17, 2009)

—— *'Xiang cai xian guan' gaige de shehuixue fenxi. Yi Anhui Sheng G Xian wei li* [A sociological analysis of the 'county administers township finances' reform: a case study of Anhui's G County], unpublished Master's thesis, Anhui University, 2005

Zhou, Z., 'The peasant burden has multiple causes', *Fazhi Ribao*, May 21, 1996, p. 7, FBIS-CHI-96-152

Zhu, A. and Yao, C., 'Nongcun shuifei gaige shidian yixian zhuizong (yi)' [Tracing the path of the RTFR , Part 1], *Zhongguo caijing bao*, October 13, 2003

Zhu, G., *Xiangcun zaiwu guimo yu zaiwu fengxian* [The extent and dangers of rural debt], manuscript, Beijing, 2006

Zhu, R., 'Jiu jie renda sici huiyi juxing jizhe zhaodaihui. Zhu Rongji zongli da Zhong wai jizhe wen', *Renmin Ribao*, 16 March 2001. Online. Available HTTP: http://www.people.com.cn/GB/shizheng/16/20010316/418508.html (accessed on November 1, 2009)

Zhu, Rongji, *Zhu Rongji zongli zai Guowuyuan di yi ci quanti huihi shang de jianghua zhaiyao* [A summary of Premier Zhu Rongji's speech at the first full meeting of the State Council], March 24, 1998, internal reference, 1998

Zhu, S., Zhang, H. and Yan, H., 'Nongcun shuifei gaige shidian he xiangcun guanli tizhi gaige genzong yanjiu baogao' [Follow-up research report on the Rural Tax and Fee Reform trial areas and rural administrative system reforms], *Jingji Yanjiu Cankao* 2003 (40), 2003, pp. 2–24

Index